# Alaska Dinosaurs
## An Ancient Arctic World

# Alaska Dinosaurs
## An Ancient Arctic World

**Anthony R. Fiorillo**
**Perot Museum of Nature and Science**
**Dallas, Texas, USA**

CRC Press
Taylor & Francis Group
Boca Raton London New York

CRC Press is an imprint of the
Taylor & Francis Group, an **informa** business

Dinosaur-bearing rocks in Aniakchak National Monument, Alaska.

CRC Press
Taylor & Francis Group
6000 Broken Sound Parkway NW, Suite 300
Boca Raton, FL 33487-2742

First issued in paperback 2020

© 2018 by Taylor & Francis Group, LLC
CRC Press is an imprint of Taylor & Francis Group, an Informa business

No claim to original U.S. Government works

ISBN 13: 978-0-367-65744-4 (pbk)
ISBN 13: 978-1-138-06087-6 (hbk)

### Library of Congress Cataloging-in-Publication Data

Names: Fiorillo, Anthony R., author.
Title: Alaska dinosaurs : an ancient Arctic world / Anthony R. Fiorillo.
Description: Boca Raton : Taylor & Francis, 2018. | Includes bibliographical references.
Identifiers: LCCN 2017030286 | ISBN 9781138060876 (hardback : alk. paper)
Subjects: LCSH: Dinosaurs--Alaska. | Paleontology. | Alaska--Antiquities.
Classification: LCC QE861.8.A4 F56 2018 | DDC 567.909798--dc23
LC record available at https://lccn.loc.gov/2017030286

**Visit the Taylor & Francis Web site at**
**http://www.taylorandfrancis.com**

**and the CRC Press Web site at**
**http://www.crcpress.com**

# Dedication

*To Charles Repenning, the vertebrate paleontologist who first identified dinosaur bones from Alaska*

# Contents

# Preface

The perfect journey is never finished, the goal is always just across the next river, round the shoulder of the next mountain. There is always one more track to follow, one more mirage to explore.

**Rosita Forbes, English travel writer and explorer**

There is only one means by which to understand the history of life on Earth, and that is through the fossil record. Public fascination with fossils is historical, dating back at least to the earliest public displays of dinosaurs in the mid- and late-19th century. The discovery of an abundance of dinosaur remains in western North America in the mid- to late-19th century can be said to have given the United States its first unique scientific contribution to the global community. That this fascination with paleontology, and particularly dinosaurs, continues today is clear from the vast number of stories in the news regarding fossils. This fascination stems from the simple fact that fossils help us understand today's world: they inform us about how we got here, and they contribute to telling us where we might be going.

Fossils tell an extraordinary story, and there is no shortage of books available to tell those stories. But I hope that this book reads as more than just another dinosaur book. An abundance of books are available recounting heroic exploration adventures of the past and man's constant battle against nature. These stories tell of a primal drive for survival against overwhelming odds generated by bad planning, bad luck, or a combination of both. Often the more popular of these books are stories in which some or all of the members of an expedition didn't survive the ordeal. Fortunately, this work doesn't follow that path.

This work is mostly set in and near the Arctic, a place that conjures up feelings of romance and danger. In contrast to the harrowing stories of yesteryear, the search for Alaskan dinosaurs has occurred in what might affectionately be referred to as "the banana belt of the Arctic," so called because the work occurred during the relatively calm and warmer summer months. Of the now thousands of photographs taken over the course of this project, so many show a beautiful blue sky as the backdrop. These photographs could lead one to think this exercise was like a tranquil summer vacation; when the weather was not cooperating, in fact we were often busy trying to stay warm and dry.

However, despite this relatively benign fieldwork compared to the explorations of the past, part of the excitement of fieldwork in the high latitudes is the opportunity to encounter the unexpected and be challenged by what you think you know. That is the aspect that draws individuals more intrigued by wandering, often alone, in remote locales rather than sitting at a desk in an office. Life in field camps was often so stimulating and vigorous as to be breathtaking. However, while some of us preferred the field camp setting, the successes reaped during those times benefited from the support and teamwork of those who preferred working in the office and those who supported us financially. Those individuals bought into a vision of why this work mattered, so their logistical and financial contributions proved equally valuable.

The naturalist John Burroughs confessed in his preface to the collection of essays in *Time and Change* (1912) that his interest in things geologic exceeded his knowledge. Similarly, it has become clear over the years that my interest in Alaska, and the rest of the Arctic, greatly exceeds my knowledge.

## REFERENCE

Burroughs, J. 1912. *Time and Change*. Boston, MA: Houghton Mifflin Company.

# Acknowledgments

What we think we now know about Arctic dinosaurs took the help of many people, including those who engaged in the scientific issues surrounding exploration, those who provided logistical support, those who funded the work, and those who just followed the various projects and would stop and ask how things were going when they would see me in public places like the Denali National Park Visitor Center complex. This work has benefited from the many who shared the vision.

The National Park Service has been a great friend to this work, and I thank Vincent Santucci for making early introductions for me and providing some much-needed guidance regarding the operations of the National Park Service. Much of this story would have remained untold had it not been for people from the Alaska Region of the National Park Service. While he could have easily dismissed a guy from Dallas trying to work in Alaska as a boondoggle to escape the summer heat of Texas, I thank Russell Kucinski (now retired) for buying in early to the vision for paleontological surveys before any dinosaur had been found in any national park unit in Alaska. Russ offered his support and connected me to additional people who might make the vision a reality.

Among the National Park Service present and former staff, I thank Peter Armato, Paul Anderson, Philip Hooge, Guy Adema, Amanda Austin, Chad Hults, Dale Vinson, Sarah Venator, Lucy Tyrrell, Denny Capps, and Troy Hamon. Each offered support and guidance based on their experiences, and those experiences contributed to many of our successes.

I particularly thank Linda Stromquist, now retired from the National Park Service, as she took on the role of paleontology coordinator for much of the run of this work in the national parks in Alaska. Her efforts were tireless, and she was able to get the projects past bureaucratic hurdles that seemed insurmountable to someone who worked outside the agency. She was also instrumental in helping secure the needed funding to keep this project moving forward. Linda was also a valuable colleague on many of the field projects along the Yukon River, where after many attempts we ultimately achieved success. Her dedication was a key factor to our success within the Alaskan National Parks.

I also gratefully acknowledge the efforts of Phil Brease, former geologist at Denali National Park, as well as his wife Barb, for their enthusiasm in the early planning of the work in that park as well as being so incredibly welcoming as the project unfolded. Phil's experience, patience, and vision helped shape what we ultimately learned. Phil also helped secure early funding for the work in Denali National Park discussed here. Phil's willing introductions to other geologic colleagues led to fruitful conversations with Jeffrey Trop and Kenneth Ridgway. Jeff and Ken were extremely generous with their thoughts on where to start the search for dinosaurs in Denali National Park, and that generosity certainly helped our subsequent success.

NJ Gates and David Tomeo of Alaska Geographic were also enthusiastic supporters of this work, including their annual invitation for me to speak at the Murie Science and Learning Center in Denali National Park. And of course, David's keen eye brought to light the discovery that placed therizinosaurs, those highly unusual dinosaurs, in the Cretaceous of Alaska. Interactions with the staff at the Murie Science and Learning Center, the staff at Denali National Park, the park bus drivers, and the visiting public were always something to look forward to.

In addition to joyfully enduring various aspects of camp life, other colleagues provided stimulating conversations throughout the time spent in the field camps. Particularly, I thank Thomas Adams, Brent Breithaupt, Michaela Contessi, Federico Fanti, Erin Fitzgerald, Peter Flaig, Roland Gangloff, Steve Hasiotis, Louis Jacobs, Yuong-Nam Lee, Amaury Michel, Kent Newman, Judith Parrish, Jason Petula, Peter Rose, Paul Sereno, Chris Stragnac, Tomonori Tanaka, Carla (Susi) Tomsich, Dolores Van der Kolk, Ryuji Takasaki, Grant Shimer, and Lisa Zago for their engaging company. Additional conversations with colleagues such as Greg Ludvigson, Larry Hinzman, and Celina Suarez expanded my horizons as well. With his interest in Arctic science, Mead Treadwell helped facilitate this work through his seemingly endless connections to a wide variety of communities, both in Alaska and beyond.

I thank Louie Marincovich, Jr. for his permission to include his previously unpublished account of the series of events leading to the recognition of dinosaur bones from the Prince Creek Formation along the Colville River. I also thank Louis Taylor for arranging my initial meeting with Charles Repenning, and of course I thank Charles Repenning for his generosity in sharing his early thoughts on the recognition of Arctic dinosaurs from Alaska.

Beth Hook, formerly of the Perot Museum of Nature and Science, was instrumental in creating a strong presence of this work for the public. I also thank Kimberly Jones and the Dallas Museum of Art for permission to reproduce *The Icebergs* for this book. The responsibility for managing the collections made during this project fell to the capable hands of Karen Morton, Collections Manager for the Perot Museum of Nature and Science.

This work did not occur by magic or by sheer will—it was accomplished through the generosity and partnership of several sources of financial support. The Arctic is an expensive place in which to operate, and funding can be very difficult to obtain. The funding provided by the following individuals and organizations was critical to our successes in the field. The members of the Perot Paleo Club (Lane and Kate Britain, Don and Cheryl Coney, Scott and Kelly Drablos, Dick Hart, Dick and Sheryl Latham, Tom and Sharon Meurer, Suzy Ruff, Danny and Judith Tobey, Robert and Sharon Van Cleave, Michael and Gretchen Vick, Alex and Jacqui Winslow) provided much-needed assistance during phases of this work that allowed it to continue. I especially thank Ron and Jane Gard, Cathey and Donald Humphreys, and Virginia and Ansel Condray, the founding members of the Perot Paleo Club, for their early support of this project. Additional funding for this work came through a variety of additional sources including the National Park Service, the National Science Foundation, the National Geographic Society, the Explorers Club, the Vick Charitable Foundation, the Ocean Alaska Science and Learning Center, the Jurassic Foundation, and the Foundation Mamont.

Ron Tykoski's skills in the fossil preparation lab at the Perot Museum of Nature and Science as well as in graphic design of many of the figures in this book have been a huge asset. His engaging discussions and commitment to the fossils brought back to the lab have helped shape our understanding of biodiversity in the ancient north. His efforts in the lab matched what others did in the field.

I also thank Dave Norton for providing so much guidance throughout my own personal journey of exploring for dinosaurs in Alaska. Dave's excitement and curiosity about the North seems to be boundless, and his patience for a "cheechako" equally endless. From understanding aspects of the natural history and human history of the Arctic, to reading rivers and learning how to drive boats, to watching the weather or birds, or almost anything, I quickly learned that if Dave started a sentence with "Isn't this interesting ..." I should stop what I was doing and listen.

My colleagues Yoshitsugu Kobayashi and Paul McCarthy have been invaluable throughout so many of the successes leading to this book. The chemistry created through the companionship and exchange of ideas with these two modest yet incomparable colleagues has been a magical experience—one that I would hope aspiring scientists might be able to emulate with colleagues of their own at some point in their careers.

While this book may be imperfect, it is less imperfect thanks to the review comments provided by Louis Jacobs, Dave Norton, Paul McCarthy, Ron Tykoski, and Bonnie Jacobs. I thank each of them for their willingness to review parts or all of this manuscript as it was being prepared. I also thank Chuck Crumly for his encouragement to write this book.

And last, I thank all the members of my family, particularly my wife Jessica and our daughter Olivia, for their patience over the course of this long project. I don't think anyone expected the project to continue for as long as it has. Without their support, I would not have been able to do this work.

# Author

**Anthony R. Fiorillo, PhD,** earned a BSc from the University of Connecticut and an MSc from the University of Nebraska. He completed his PhD in vertebrate paleontology, specifically dinosaur taphonomy and paleoecology, at the University of Pennsylvania, became a Carnegie Museum of Natural History Rea Postdoctoral Fellow, and then a museum scientist at the Museum of Paleontology at the University of California–Berkeley. Tony joined the Museum in 1995 and he is now the vice president of Research & Collections and chief curator of the Perot Museum of Nature and Science.

Tony also holds adjunct positions with Southern Methodist University in the Roy M. Huffington Department of Earth Sciences, the University of Alaska Department of Geosciences, and the Hokkaido University Museum. He chaired, co-chaired, and served on the Education Committee of the Society of Vertebrate Paleontology from 1996 to 2004. His work on ancient terrestrial ecosystems has taken him into over a dozen units of the National Park Service and won him national recognition in 2007, receiving the National Park Service, Alaska Region Natural Resource Research Award. He became a Geological Society of America Fellow in 2008 and in 2013 he was bestowed the Distinguished Alumni award by the University of Nebraska Department of Earth and Atmospheric Sciences. For him, the appeal of the work in Alaska is the result of the combination of intellectual pursuit and the rigors of working in the Arctic environment. Tony has published over 100 scientific and popular papers, as well as co-edited four scholarly volumes.

# 1 The Age of the Ancient Arctic

My first sighting of a polar bear in the wild took place under nearly ideal conditions. Colleagues and I had just finished my first field season in the Arctic digging up dinosaur bones, and we were driving our boat down the Colville River back to the native village of Nuiqsut. We had just rounded a bend in the river when we came upon a startled white bear, so from the comfort of my noisy boat, I watched as the bear frantically swam away from us. We were at the time several tens of kilometers upriver from the Beaufort Sea coast so not only was our intrusion not welcomed by the bear, our stumbling across this bear was entirely unexpected, so much so that at the time we reported our bear encounter we were told that it was only the second time anyone had ever seen a polar bear that far inland during the summer. For nearly a month we had been encountering car-sized wedges of ice eroding out along some of the river banks, walking the uneven patterned ground controlled by the underlying permanently frozen ground, tripping over tundra vegetation that only reached to our ankles, adjusting to sleeping in the endless daylight, and feeding the multitude of ever-present mosquitoes. Of all these experiences with aspects of the North, encountering this polar bear came to represent to me the quintessential epiphany that I had indeed spent that month in the Arctic.

It is hardly profound to declare that the Arctic is a wondrous and inspirational place. While there may be other places to camp where one can be roused from a deep sleep to find oneself staring into the eyes of a grizzly bear, the Arctic is a place not only where that can happen, but also one of only two places where collecting fossils could involve chipping through rock permeated with ice (the other place of course being Antarctica). And while there may be other places on Earth where the sound of insects is persistent, the Arctic offers that cacophony 24 hours a day—thanks to the nearly endless daylight. With its short summers, the Arctic is also a place where one can pitch a tent in early July and watch flowers bloom, only to fold up camp by early August having watched those same flowering plants go to seed. While living among these natural wonders, it also becomes clear to an Arctic visitor that when one goes out to experience these and other natural marvels, one quickly goes "off the grid" such that independence of a form quite different from the independence cultivated while living in a more urban or community setting is called into play.

Seeing the Arctic as a wondrous and inspirational place reflects a certain philosophy and spirituality. Dave Harmon (2016) suggests that the reason for the popularity of the United States National Park Service lands is that visitors are exposed to incomparable scenery, and the immersive experience with the scenery creates a sense of wonder and inspiration, a sense of place. The participating visitor is challenged to understand something about the landscape, and through meeting these challenges can reach some level of fulfillment (Harmon, 2016).

A review of the body of literature and art dedicated to the Arctic supports a similar sense of place, or the notion of a connection to the inspirational aspects of the region. This connection is so evident that even those from the objective sciences, such as pragmatic mathematical ecology, can articulate an appreciation (Pielou, 1994). This association of inspiration and art connects that sense of wonder to the broad idea of the Sublime, a conceptual framework that can be traced back to the ancient Western world where gods and heroes raised the human consciousness closer to the divine (Wilton, 2002).

Originally stated within the context of the body of work of explorer-artists from the 19th century and earlier, this mystical mix of feelings has been referred to as the Arctic sublime (Loomis, 1977). The spiritual nature of the Arctic is perhaps best known to many through the literary efforts of Barry Lopez's *Arctic Dreams* (1986), and perhaps through the art of Frederic Edwin Church's painting *The Icebergs* (Figure 1.1). That famous painting of the Arctic resides at the Dallas Museum of Art. Frederic Edwin Church was part of the Hudson River School of American landscape artists, and based on his experiences during a voyage to the North Atlantic in 1859, Church produced *The Icebergs* in the early 1860s. The painting is substantial in size and depicts an even more substantial landscape, that of an oceanic perspective of the Arctic in eastern North America. The artwork is dominated by the interplay of light and ice in the image portrayed; in the foreground, there is the broken mast of a wooden sailing ship. The concept of the Arctic remains foreign to many people today, but it was arguably even more so in the 1860s when this painting was unveiled. The

**FIGURE 1.1**    Frederic Edwin Church, *The Icebergs* (1861). Oil on canvas. Image dimensions: 64.5 × 112.5 in. (1 m; 63.83 cm × 2 m 85.751 cm). Framed dimensions: 85 × 133 × 5 in. (2 m 15.9 cm × 3 m; 37.821 cm × 12.7 cm). Weight: 425 lb. (192.78 kg). Dallas Museum of Art, gift of Norma and Lamar Hunt. (Image courtesy from Dallas Museum of Art, 1979.28. With permission.)

history of this famous painting, its conceptual evolution, its political implications, and its long period of being "lost" are arguably as interesting as the landscape it portrayed (Lopez, 1986; Harvey and Carr, 2002). That broken mast of the wooden sailing ship speaks to the almost incomprehensible scale of the Arctic to those more familiar with the lower latitudes. In the original unveiling of the work, the painting was only ice and light, but the public reaction was so tepid that the broken mast was added to provide a more identifiably human sense of scale (Lopez, 1986; Harvey and Carr, 2002).

The luminism of the work is one of the most interesting parts of the painting. The light appears to be coming from somewhere off to the left, outside of the scope of the painting. The perspective supposedly depicts late afternoon light (Harvey and Carr, 2002), which means viewers of the painting are facing north. One could interpret the setting as, with the daylight fading, that by looking northward, the future, is to the north.

Much as some people are drawn to deserts, some are drawn to the polar regions. Both of these extremes can fall within two of the five classic zones of the ancient Greeks, the Burning Zone and the Frigid Zone, respectively (Stefansson, 1944). It was the astronomical observations of the ancient Greeks that led to an understanding of the Arctic Circle (Mirsky, 1948). The recorded exploration of the Arctic begins with the Greeks and the Vikings (Mirsky, 1948). Driven by curiosity, commercial gain, sovereignty, adventure, and knowledge, these early explorations set the back-drop for the lure of the Arctic, which led to the zenith of heroic, larger-than-life Arctic exploration in the 19th century.

In contrast to this conquering cast of Arctic heroes that flourished in the 19th century and then spilled over into the early 20th century, that epically pitted man against nature, Stefansson (1921) recognized that rather than a barren wasteland filled with constant life or death struggles, native peoples had thrived in the Arctic for millennia and he termed the region "The Friendly Arctic." Westerners' cultural biases had long blinded Western explorers to adaptive indigenous strategies and pro-hibited them from adopting what measures it took to flourish in the Arctic. The Arctic sublime, then, is a specific cultural notion that implies that the Arctic holds some physical or psychological mysteries connected to the human psyche and that in-depth understanding will allow humans to appreciate a renewed sense of purpose with the land (Oshenko and Young, 2005).

Nevertheless, as Stefansson (1921) foresaw, in addition to the unique attributes of the Arctic environment, the Arctic also holds such great strategic importance in geo-political terms. The region's unique importance has continued to grow since then, making some suggest that Arctic nations have more in common than what divides them (Ramseur, 2017). Advances in technology and travel have resulted in increased access to the region causing some to consider that the world is on the threshold of the "Age of the Arctic" (Young, 1985; Osherenko and Young, 2005). While the glaciologist Carl Benson once remarked that "ignorance of the Arctic is an infinite resource" (Norton, 2011), the need to learn more about all aspects of the region is increasingly—even urgently—evident. It has become clear that changes in global climate are most profoundly expressed in the polar regions. Biological consequences aside for the moment, symptoms of a warming globe include the simple fact that feasible shipping lanes within the Arctic now extend farther spatially and last longer seasonally because sea ice blocking the way for ships melts earlier in the spring and refreezes later in the autumn. Several nations now make conflicting claims to sea-floor sectors and islands within the Arctic Circle. In light of potential international confrontations, understanding the nature of Arctic geologic, meteorological, oceanic, and atmospheric phenomena assumes critical importance. How these phenomena may affect or constrain the surveillance or weapons technology is relevant and significant (Young, 1986). Thus, some students of the Arctic have argued that we need to recognize the Arctic as not only a well-defined circumpolar geographic region but also as one with growing socioeconomic and geopolitical importance (Young, 1986; Osherenko and Young, 2005; Anderson, 2009; Strusik, 2009; Ramseur, 2017).

The Arctic is centered on latitude 90° north, or the geographic North Pole. Ice and a general lack of trees dominate terrestrial biomes at higher latitudes, and permanently frozen ground is present throughout much of the Arctic region. Arctic climate is typically cold and dry, but given the sheer size of the Arctic region and its heterogeneous topography, climate can be highly variable, yielding several distinct ecosystems. Solar energy is lower in the Arctic than in the lower latitudes as a simple function of geometry. Rays from the sun that strike the Arctic must pass a longer distance through the atmosphere and then are spread over a wider area on the surface of the Earth than the rays that strike the lower latitudes (Figure 1.2). The result is that more energy is reflected or absorbed by the atmosphere in the Arctic than in the more equatorial regions.

Biologists have long recognized a general trend of reduction in species diversity moving from the lower latitudes to the higher latitudes, and that pattern holds even within the Arctic region (Figure 1.3). This species reduction with increasing latitude is likely the result of the sharply defined seasonal windows of ecosystem productivity that result from the marked annual peaks and troughs in thermal energy and photoperiod.

With respect to terrestrial biota native to the modern Arctic, the flora contains approximately 1500 species of vascular plants, 1200 types of lichens, and 750 types of bryophytes (Walker, 1995), compared to two-thirds of all plants that live in tropical rainforests. It is commonly recognized that these Arctic plants do the same type

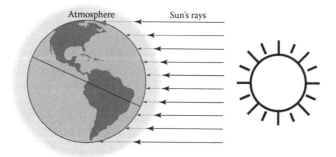

**FIGURE 1.2**  The relationship of energy passing through the atmosphere at high latitude compared to low latitude. Notice that the distance energy passes through the atmosphere is higher in the higher latitudes than in the lower latitudes, and energy is dispersed over a greater area in the higher latitudes.

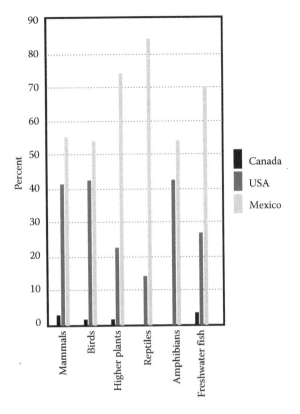

**FIGURE 1.3** Comparison of species diversity in Canada, the United States, and Mexico as a means of showing decreased diversity within increased latitude. (Redrawn from Pisanty-Baruch, I. et al. 1999. *The George Wright Forum* 16:22–36.)

of things that plants in the lower latitudes do: they fix and sequester carbon by photosynthesis and anabolism, they respire, and they absorb nutrients. Blix (2005) asked if there is anything special in evolutionary terms that characterizes Arctic plants. His suggestion was that it is the sum of adaptations of these plants living in such an extreme environment that provides them the ability to accomplish basic component processes of plant physiology and makes these plants successful in the Arctic (Blix, 2005).

Successful terrestrial fauna in the Arctic copes with a set of challenges somewhat different from those facing plant life. The Arctic features long, dark nights in the winter months and long days of sunlight during the summer months. The reason for this lies in the astronomical matter of the tilt of the Earth's axis of rotation with respect to the plane of its orbit around the sun. This tilted axis maximally inclines the Earth's Northern Hemisphere toward the sun during June and away from sun in December (Figure 1.4). On summer solstice (June 21) all of the area within the Arctic Circle—that line corresponding to approximately latitude 66° and 33 minutes north—is exposed to 24 hours of sunlight for the duration of one day. The size of the area contained within the Arctic Circle is approximately 4% of the entire

**FIGURE 1.4**   Schematic drawing showing how the axis of the Earth is inclined such that the Earth's Northern Hemisphere is tilted toward the sun during June and away from the sun in December.

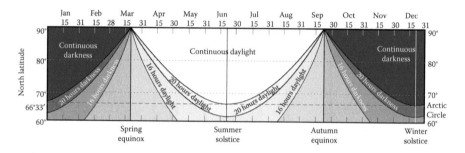

**FIGURE 1.5**   The distribution of daylight and darkness as a function of latitude and calendar year.

surface of the planet. Conversely, on winter solstice (December 21) all of that area within the Arctic Circle is beyond the reach of direct sunlight (Figure 1.5). While this seems simple, the Arctic Circle of today is not necessarily the Arctic Circle of yesterday. Due to the wobble of the Earth caused by the orbit of the moon, there is approximately a 2° variation in the latitude of the Arctic Circle over the course of its 40,000-year cycle of wobble in the angle of the Earth's axis. Along with this pattern of photoperiod extremes, another unique component of the polar environment is that in the months when the days are experiencing both light and dark, the changes in length of those two extremes can change by as much as 30 minutes each day (Blix, 2005). It should also be noted that if one stood at sea level at 90° north on March 21–22, the sun rises only once before setting again on September 21–22. Thus, day length increases not by 30 minutes each day, but instantaneously by 24 hours per day. Such dramatic changes in photoperiod are unique to the polar regions, resulting in one of the most extraordinary adaptations by animals to a high-latitude lifestyle— the decoupling of the circadian rhythms found in lower latitude fauna. The modern Arctic biota presents marvelous opportunities to study the resilience of life by virtue of its existence in an extraordinary and extreme environment.

In the following chapters, I hope to establish that studies of polar dinosaurs, while presenting daunting physical challenges, are more than a means to test one's mettle, and more than reruns of the 19th-century heroic polar exploration model. Although high-latitude fieldwork can present enormous physical and mental challenges beyond the collection of scientific data, studying the Arctic in deep geologic time shows that the Arctic of the Cretaceous was much warmer than it is today, and that Arctic ecosystem(s) were biologically richer in terms of both abundance and diversity.

While a number of important Mesozoic terrestrial vertebrate fossil localities now exist within the northern high latitudes (notably in Canada, Norway, Russia, and Sweden), the growing fossil record from the Cretaceous rocks within just the state of Alaska has provided a robust data set, furnishing us with insights into ancient Arctic and sub-Arctic terrestrial ecosystems. During the latest part of the Cretaceous, herds of plant-eating duck-billed dinosaurs roamed through varied landscapes from the far northern part of what is now modern Alaska southward to the Pacific Rim coasts of the state. Herds included not only the adults but also age classes from the very young to the nearly full-grown individuals. This biogeographic pattern of herd distribution and composition suggests that the ancient Arctic consisted of dynamic, varied, and productive ecosystems.

Along with the duck-billed dinosaurs that seemed to dominate the landscape, there were also herds of plant-eating horned dinosaurs, specifically *Pachyrhinosaurus*, a cousin of the iconic *Triceratops*. And as one would expect in well-developed ecosystems, there were also animals that functioned as predators on herbivores, such as the tyrannosaur *Nanuqsaurus* and the smaller meat-eating dinosaur *Troodon*. Given the rigors of the physical environment of the ancient north, these higher trophic level (i.e., meat-eating) dinosaurs seem to have been well adapted for their role in the Cretaceous Arctic.

We are just beginning to appreciate the diversity of landscapes through which these and other animals roamed. In the far north, for example, the polar woodlands of the Late Cretaceous were dominated by gymnosperms (conifers and allies) with an angiosperm (flowering plants) understory. This region of Alaska features several fossil bonebeds consisting of many individual herbivorous dinosaurs whose apparently simultaneous deaths raise the question of most likely candidates for a killing mechanism. This ancient Arctic coastal plain was influenced by seasonally varying hydrologic processes, water levels, and by what appear to have been episodic floods. It is easy to imagine severe spring runoff events due to the paleogeographic juxtaposition of the towering ancestral Brooks Range, a narrow coastal plain, and the Arctic Ocean of the Cretaceous. Spring melting of the high-altitude snowpack may sufficiently explain some or all of these bonebeds (Fiorillo et al., 2010).

Elsewhere in the region, the vegetative cover was also dominated by conifers with flowering plants taking minor roles in either the forest understory or along rivers, but the openness of the vegetative cover varied, as did the specific community composition of various biotic communities. Sedimentation rates seem to have varied across the region as well.

Taken as a whole, then, Cretaceous Alaska emerges as a diverse region. It is tempting to imagine today's many visitors to National Park Service units in Alaska having their wonder at modern scenic and natural history amplified by our growing understanding of Cretaceous paleoecology, and specifically dinosaurian ecosystems.

Global climate change is a current topic of intense concern. Given that such change profoundly affects the biological world, the importance of understanding climate change is growing. In the absence of a modern analog for a warm Arctic, many aspects of the climate change discussion are concerned with predicting the changes brought about by warming of the polar region. Therefore, understanding Cretaceous Arctic ecosystems should provide indirect evidence from deep time when a much warmer Earth existed under a light regime similar to the present one. To understand the context of the modern world, one must examine how we got here, and what follows from this growing number of studies is the demonstration of the rich ecosystems that thrived in this ancient warm polar world. Such a synthesis provides a new and widened perspective on Arctic environments.

The contents of this book should not be viewed as a closing milestone—the equivalent of the setting sun in the painting. Instead, the following chapters may better serve to introduce readers to what has remained until now a largely untapped moment in geologic history. In addition to challenging much of what we think we know about dinosaurs, looking back in deep geologic time, this work purports to show that the ancient Arctic is as wondrous and inspirational as many have discovered in its modern counterpart. By pursuing understanding of the growing significance of an ancient Arctic, it may be that we have now entered an expanded definition of the "Age of the Arctic," or perhaps more precisely, "The Age of the Ancient Arctic."

## REFERENCES

Anderson, A. 2009. *After the Ice: Life, Death and Geopolitics in the New Arctic.* New York, NY: HarperCollins Publishers.

Blix, A.S. 2005. *Arctic Animals and Their Adaptation to Life on the Edge.* Trondheim: Tapir Academic Press.

Fiorillo, A.R., P.J. McCarthy, and P.P. Flaig. 2010. Taphonomic and sedimentologic interpretations of the dinosaur-bearing Upper Cretaceous Strata of the Prince Creek Formation, Northern Alaska: Insights from an ancient high-latitude terrestrial ecosystem. *Palaeogeography, Palaeoclimatology, Palaeoecology* 295:376–388.

Harmon, D. 2016. Sense of place. In *A Thinking Person's Guide to America's National Parks,* eds. R. Manning, R. Diamant, N. Mitchell, and D. Harmon, 21–29. New York, NY: George Braziller Publishers.

Harvey, E.J. and G.L. Carr. 2002. *The Voyage of the Icebergs: Frederic Church's Arctic Masterpiece.* New Haven, CT: Dallas Museum of Art and Yale University Press.

Lopez, B. 1986. *Arctic Dreams.* New York, NY: Charles Scribner & Sons.

Loomis, C.C. 1977. The arctic sublime. In *Nature and the Victorian Imagination,* eds. U.C. Knoepflmacher and G.B. Tennyson, 95–112. Berkeley, CA: University of California Press.

Mirsky, J. 1948. *To the Arctic: The Story of Northern Exploration from Earliest Times to the Present.* New York, NY: Alfred A. Knopf.

Norton, D.W. 2011. Review of Siku: Knowing our ice. *Documenting Inuit Sea-Ice Knowledge and Use,* eds. I. Krupnik, C. Aporta, S. Gearheard, G.J. Laidler, and L.K. Holm, 501pp. Dordrecht, Germany: Springer, 2010. *Arctic* 64:381–384.

Osherenko, G. and O.R. Young. 2005. *The Age of the Arctic: Hot Conflicts and Cold Realities.* Cambridge, UK: Cambridge University Press.

Pielou, E.C. 1994. *A Naturalist's Guide to the Arctic.* Chicago, IL: University of Chicago Press.

Pisanty-Baruch, I., J. Barr, E.B. Wiken, and D.A. Gauthier. 1999. Reporting on North America: Continental connections. *The George Wright Forum* 16:22–36.

Ramseur, D. 2017. *Melting the Ice Curtain.* Fairbanks, AK: University of Alaska Press.

Stefansson, V. 1921. *The Friendly Arctic: The Story of Five Years in Polar Regions.* New York, NY: The Macmillan Company.

Stefansson, V. 1944. *Arctic Manual.* New York, NY: The Macmillan Company.

Strusik, E. 2009. *The Big Thaw: Trends in the Melting North.* Mississauga, ON: John Wiley & Sons.

Walker, M.D. 1995. Patterns and causes of Arctic plant community diversity. In *Arctic and Alpine Biodiversity*, eds. F.S. Chapin and C. Korner, 3–20. Berlin, Germany: Springer-Verlag.

Wilton, A. 2002. The sublime in the old world and the new. In *American Sublime: Landscape Painting in the United States 1820–1880*, eds. A. Wilton and T. Barringer, 11–37. Princeton, NJ: Princeton University Press.

Young, O.R. 1985. The age of the Arctic. *Foreign Policy* 61:160–179.

# 2 History Leading to Arctic Alaska Dinosaur Discoveries

The generally accepted earliest evidence of humans in the Arctic, migrants from eastern Asia, dates to approximately 13,000–14,000 years ago (e.g., Hoffecker et al., 1993), though newer discoveries have suggested the age could be doubled to approximately 27,000 years ago (Pitulko et al., 2004). By contrast, record-keeping navigators and sailors setting out from circum-Mediterranean agricultural settlements only began tentatively to probe northward after several thousand years had passed.

During the time Alexander the Great was marching toward India in the 4th century BCE, Pytheas, a Greek mathematician and astronomer from Massilia (now Marseilles) in southern France, was sent north to find the remote sources of the trade goods tin and amber. His actual travel records are lost, but stories of his explorations and discoveries seemed so outlandish that when referenced years later by Classic Greek historians such as Polybius and Strabo, the travels and marvels discovered were viewed with skepticism. But from these early historians references to Pytheas' travels we get the term *Ultima Thule*, or "beyond the borders of the known world."

Massilians were considered good sailors of the day, so an epic voyage is not beyond reason, and it is thought that Pytheas essentially started his expedition following the path set by an even earlier explorer, Himilco, a Carthaginian who sailed some aspect of the coast of the west or northwest part of Europe. The farthest limit of Himilco's exploration is unclear but may have been what is now northwestern France, or perhaps only as far as Cape Finisterre, a point of land on the west coast of Spain, a name that in Latin literally means "end of the Earth."

Pytheas sailed out through the Pillars of Hercules, known today as the Strait of Gibraltar, and followed the western coast of Europe and then continued north. The terminus for this voyage is unclear but he is credited with being the first to record the Midnight Sun and polar ice, thus earning him the reputation of being the first scientific explorer. The early skepticism that greeted Pytheas' exploits proved to be a foretaste of the future. Descriptions of the Arctic world that explorers would bring back to those in more temperate regions challenged and excited human imaginations with seemingly "other-worldly" accounts.

Jeannette Mirsky (1948) chronicles the accelerating pace of western exploratory probes into the Far North over the 24 centuries following Pytheas' search for *Ultima Thule*, while Hector Chevigny (1965) recounts the history of the Russian American period in Alaska, AD 1741–1867. Although both books are informative and entertaining, neither of these chroniclers fully explores the public receptions that greeted the return of the many expeditions to their ports of origin, which had been initiated variously by economic gain, personal glory, altruistic rescue of other explorers, or nationalism. Nor could these authors do justice to some of the more recent advances in scientific understanding of global dynamics contributed through observations of modern Arctic processes and systems. Examples of these include the milestones of sea ice-based drifting research stations begun by the USSR in 1937 and continued by Russian and U.S. scientists after World War II (Frolov et al., 2005; Althoff, 2007), and the development of multidisciplinary scientific research at the U.S. Navy's Arctic Research Laboratory at Barrow, Alaska (Norton, 2001). Throughout most of these past 24 centuries, Arctic discoveries have generated varying degrees of skepticism and wonder at more populous latitudes—reactions that can be advantageous, as we shall see.

Tracing scientific discovery of fossil vertebrates in Alaska within general circumpolar history draws our attention first to Siberia. Russian expansion into Siberia was driven in part by the fur trade. By the close of the 17th century, the Cossacks had reached Kamchatka. One result of this expansion was generation of an interest by Russian Emperor Peter the Great in obtaining reliable maps of these vast territories. The quest for such maps left the emperor wondering whether Asia and North America were joined as one continuous landmass. Peter the Great died before he could implement what has become known as the Great Northern Expedition, or the Second Kamchatka Expedition, but the Empresses Anna and Elizabeth put those expedition plans into practice. In charge of one of the most enormous expeditionary forces history has ever seen was Vitus Bering, a Danish explorer and cartographer in Russian service, as well as a Russian naval captain.

The expedition lasted a decade, 1733–1743, and one of the participants on the expedition was Georg Wilhelm Steller, a German scientist, responsible for collecting specimens and making scientific observations. During the crossing of the strait between Asia and North America in 1741, a storm hit the two ships under Bering's command, the *St. Peter* and the *St. Paul*. Bering and Steller were on the *St. Peter*, while Bering's deputy Aleksei Chirikov captained the *St. Paul*. The storm separated the two ships, forcing them to follow somewhat different courses. Chirikov's route took the *St. Paul* to southeastern Alaska, and in the middle of July he recorded sighting land, making him the first Russian to record seeing northwestern North America. A few days later, the *St. Peter* also reached the southeastern Alaska coast. Taking advantage of this shore after an arduous voyage, Bering landed a small party to take on fresh water but was reluctant to spend more time than needed for that purpose. After all the years of preparation, Steller was granted a mere 10 hours ashore to make his observations of the natural history of this land, newly discovered by the Western world (Ford, 1966).

One can imagine the frustration of the scientist after so many years of preparation, yet in that remarkably short window of time Steller managed to make numerous important observations allowing description and naming of such birds as a jay and an eider, both of which now bear his name, Steller's jay and Steller's eider. Even with that short window in which to work, by setting foot on Alaskan soil Steller became the first European scientist to describe the natural history of Alaska. For Commander Bering, despite his lack of rigor in facilitating detailed exploration of this new land, one of the most important geographic accomplishments of his expedition was establishment by the voyage that Asia and North America were separated by water.

Sailing for the Russian Empire in an attempt to find passage across the Arctic Ocean in the early part of the 19th century, Otto von Kotzebue, while exploring the northwest coast of North America, made reference to vertebrate fossils in Alaska in what is now known as the Kotzebue Sound area (Kotzebue, 1821). By virtue of an "unpleasant storm" which forced a landing party to search for wood for a fire to stay warm, Dr. Johann Friedrich von Eschscholtz, a physician and naturalist accompanying Kotzebue, made "a very remarkable discovery" on August 8, 1818. While drying out their wet clothes from the storm, von Eschscholtz made an "extended excursion" to find part of a broken-away bank in which he had discovered a "quantity of mammoth teeth and bones" within a cliff of ice (Figure 2.1). The cliff was estimated to be 100 feet tall, with the ice disappearing under a luxuriantly vegetated mountain slope. The location of this remarkable discovery was relatively close to the area of modern day Kotzebue, Alaska (Figure 2.2).

Approximately 10 years later, Frederick Beechey visited the area of Kotzebue's discoveries along the coast of Alaska while trying to connect with Sir John Franklin as part of a coastal survey mission for northwestern North America. While in this area described by Kotzebue, Beechey collected several fossil mammal remains that were then brought to the great British geologist Reverend William Buckland. Buckland described these remains, which included specimens attributable to elephants, muskoxen, deer, and horses, as part of the published report of Beechey's multiyear voyage (Beechey, 1831). Though they did not represent dinosaurs, these fossil mammal remains

*The Icebergs of Kotzebue Sound.*

**FIGURE 2.1**  The Icebergs of Kotzebue Sound. (Illustration from von Kotzebue, O. 1821. *A Voyage of Discovery into the South Sea and Beering's Straits.* London, UK: Longman, Hurst, Rees, Orme, and Brown, Paternoster-Row, 3 volumes.)

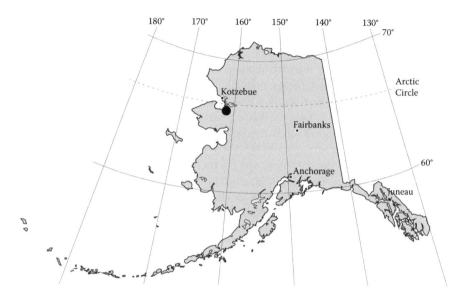

**FIGURE 2.2** Map showing the presumed location this cliff. The coordinates provided by Kotzebue (1821) indicate that the cliff was approximately 100 kilometers south of the modern town of Kotzebue, Alaska, and is presumably the basis for the name Elephant Point along Eschscholtz Bay.

provided the first scientific description of vertebrate fossils from Alaska. The discovery of the remains of dinosaurs from the polar regions would wait for over 100 years.

Dinosaurs are a historically well-defined clade (Figure 2.3), or group, of animals containing some well-known and popular species, but as time has brought new discoveries, the relationships and lifestyles of these most recognized fossil relics of the past have been active subjects of intense paleontological research. It is universally agreed that dinosaurs belong to a group known as archosaurs, which includes the modern representative crocodiles. The definition of Dinosauria is based on a number of anatomical features such as the loss of certain skull bones (postfrontals), limb and pelvic element modifications such as elongation of the deltopectoral crest on the humerus, making the upper arm bone characteristic, a brevis shelf on the ventral surface of the postacetabular part of the ilium and a well-perforated acetabulum, which make the pelvis unique, and an ascending process on the astragalus, which makes the ankle distinct from crocodiles (Benton, 2004).

Fragmentary remains of the fierce meat-eating *Megalosaurus* and the plant-eating *Iguanodon* were found in Oxfordshire, England in the 1820s (Buckland, 1824; Mantell, 1825). These bones and teeth, though fragmentary, introduced the world to a group of fossil reptiles that have come to be known as the dinosaurs (Owen, 1842). Soon after its naming, *Megalosaurus* entered classic literature in the first paragraph of Charles Dickens' *Bleak House*. The published records of dinosaur remains in North America followed a few decades later in the mid-1850s with Joseph Leidy (1856) naming four dinosaurs from Montana—*Palaeoscincus*, *Trachodon*, *Troodon*, and *Deinodon*. With the western expansion of the United States, exploration of these arid regions produced

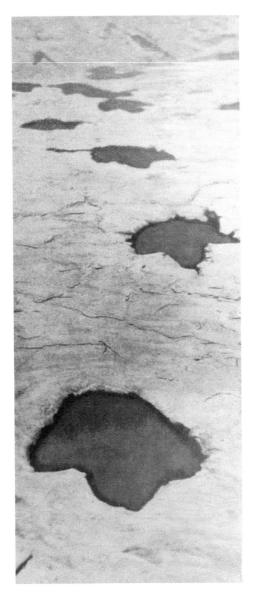

**FIGURE 2.3** This photograph, in a 1930 edition of *Encyclopédie par L'Image La Préhistoire*, appears to be ornithopod tracks from "les montagnes de l'Alaska." Given the history of energy exploration of Alaska, it is likely that the photograph was taken on the Alaska Peninsula. Thus, this photograph is the earliest known published account of a dinosaur record from Alaska.

a plethora of new dinosaur discoveries. Much has been written about the early explorers of the American West, the dinosaurs they found, and the information that was gained from their discoveries (e.g., Colbert, 1961; Rainger, 1991; Thomson, 2008).

By the late 1800s, the abundance of dinosaur bones available in the Rocky Mountains of the United States led to an explosion of information about these animals. Led primarily by Edward Drinker Cope of Philadelphia and Othniel Charles Marsh of New Haven, there was great activity in dinosaur exploration, leading to the discovery and naming of well-known dinosaurs such as *Apatosaurus, Diplodocus, Camarasaurus, Stegosaurus,* and *Allosaurus* and many others at a remarkable rate. That time has been referred to as the Great Dinosaur "Gold Rush" (Preston, 1986). A few decades later, in the early 20th century, a second great flurry of activity centered in places such as western Canada and Montana that produced a level of activity approaching that of the first great bone rush (Preston, 1986). The second bone rush introduced the world to dinosaurs such as *Tyrannosaurus rex, Styracosaurus,* and *Parasaurolophus.* The vitality of diverse dinosaur studies is arguably even more dynamic in the later 20th century and early 21st century than during either of these earlier intervals: many new taxa are being recognized from around the world at remarkable rates (Dodson, 1990; Holmes and Dodson, 1997; Wang and Dodson, 2006). Most of these newer discoveries, however, remain within the temperate to sub-temperate climatic zones of the world (Weishampel et al., 2004), continuing a significant bias in our global understanding of dinosaurs, the environments in which they lived, and thus the biogeography of their past.

In the history of dinosaur exploration, the story of Alaskan dinosaur discoveries is comparatively recent. So recent, in fact, that in 1980 the record was lampooned by Patrick McClellan (1980), a geologist by training, in the self-published article "Does the Alaskan record of land vertebrate fossils substantiate or contradict the long postulated existence of a Bering Land Bridge before the Pleistocene?" in a fictitious journal entitled *Alaskan Journal of Pre-Pleistocene Vertebrate Paleontology: The Thinnest Journal in the World.* While the paper is 13 pages long and filled with wit and parody, the abstract, which simply reads "No" and certainly qualifies as one of the shortest in history, makes the point of the brevity of the history of scientific documentation of dinosaurs from this region.

The first notice of an Alaskan dinosaur, a cryptic photograph in a 1930 edition of *Encyclopédie par L'Image La Préhistoire* (Figure 2.4), tantalizes us with what appears to be ornithopod tracks from "les montagnes de l'Alaska." It is unclear from the publication where those tracks are located, but clues come from the history of the energy industry in Alaska. Although northern Alaska now produces a significant fraction of U.S. oil, the first oil claims in the state were on the Iniskin Peninsula on the west side of Cook Inlet, southwest of Anchorage (Roderick, 1997). Upon passage of the Mineral Leasing Act of 1920 by the United States Congress, attention shifted westward in the 1920s, farther out the Alaskan Peninsula to Cold Bay. An editorial in the *Seattle Post-Intelligencer* proclaimed that the Act of 1920 had opened up the oilfields of Alaska (Roderick, 1997). With this geographic shift it seems likely that this photograph was taken in the southwestern part of the state, particularly since we now know that there are dinosaur track-bearing rock units in this area (Figure 2.5; Fiorillo and Parrish, 2004).

Aside from this 1930s photograph, most people would acknowledge that the more detailed documentation of the known record of dinosaurs in Alaska begins in the early 1960s. Dinosaur bones were first collected by Dr. Robert L. Liscomb of the Shell Oil Company from the lower Colville River in 1961 during oil exploration (Clemens, 1994; Gangloff, 2012). Tragically, Liscomb died the following year in a presumed rock slide in the southern part of Alaska, in the Prince William Sound area relatively close to

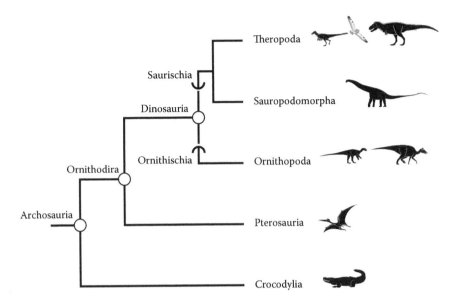

**FIGURE 2.4** Simplified cladogram showing the relationship of dinosaurs to their nearest relatives, pterosaurs and crocodiles.

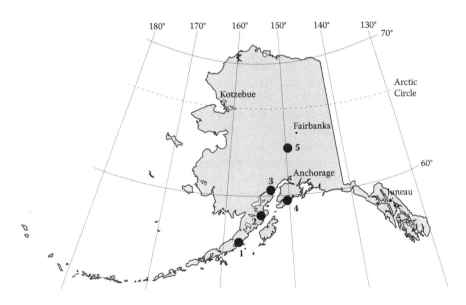

**FIGURE 2.5** Map showing the locations of the parks where the original paleontological work occurred that started the quest for dinosaurs in Alaskan National Parks. Numbers 1–5 refer respectively to Aniakchak National Monument and Preserve, Katmai National Park and Preserve, Lake Clark National Park and Preserve, Kenai Fjords National Park and Preserve, and Denali National Park and Preserve.

Cordova and far from his work along the Colville River the previous year, while conducting another geologic field project (Clemens, 1994; Gangloff, 2012). Liscomb's Colville River specimens were identified as Pleistocene, or Ice Age, mammoth bones, which are not rare in Alaska, and they remained in storage for over two decades.

Because the Liscomb discovery languished, the first technical descriptions of polar dinosaur remains can be attributed to Lapparent and his work in the Early Cretaceous of Svalbard in the Norwegian Arctic. Lapparent described a series of 13 tridactyl footprints from a sandstone slab, which he attributed to an iguanodontian dinosaur (Lapparent, 1962). This discovery, at a modern latitude 78° North, extended the northern range of known dinosaur localities almost 20° in latitude. Although the tracks did not provide particular insight into the structure of the feet of ornithopod dinosaurs, Heintz (1963) argued that the occurrence of these tracks at such high latitude suggested that these ornithopods developed some adaptation to life in the Arctic, or that Svalbard itself had shifted to its current latitude from some previous southern latitude. Although she did not express favor for one hypothesis over the other, Heintz was the first to posit the idea that high-latitude dinosaurs may have developed specialized adaptations to life in the north.

In the late 1960s, geologists working in the Upper Cretaceous Bonnet Plume Formation in the Yukon Territory of Canada discovered fragmentary remains of hadrosaurs (Rouse and Srivatasa, 1972). In a recent investigation into these exposures first studied by Rouse and Srivatasa, Evans et al. (2012) reported on fragmentary remains of non-hadrosaurian dinosaurs, thereby expanding the known dinosaurian biodiversity for the rock unit. Their study showed that the Bonnet Plume Formation holds potential for more dinosaurian discoveries, even as a review of Mesozoic-aged fossil vertebrates shows the Canadian Arctic remains a tantalizing but frustratingly incomplete record (Evans et al., 2012). As with efforts elsewhere in the Arctic, the search for dinosaurs in this potentially instructive region is hampered by logistical costs and weather (Evans et al., 2012).

Meanwhile, in the mid-1970s there were mentions by geologists performing mapping surveys in Cretaceous rocks of northern Alaska of dinosaur tracks, but these discoveries generated little interest (Campbell, 1994). Similarly, petroleum geologist R.E. Hunter discovered a slab of rock on the Alaska Peninsula that contained numerous tracks attributed to theropod dinosaurs (Campbell, 1994). These discoveries have been mentioned in popular literature but have not yet been described in a technical publication. It had been thought that these tracks from the Alaska Peninsula are Early Cretaceous in age, but Alaskan stratigrapher Frederick Wilson of the United States Geological Survey (USGS) feels strongly that the Lower Cretaceous Staniukovich Formation only crops out at the extreme end of the Alaskan Peninsula. He suggested to me in a conversation in 2000 that the tracks may be Late Jurassic instead. More recently, a joint effort led by the University of Alaska re-located this intriguing fossil locality in an effort to document it properly. These investigations support the contention that the tracks are Jurassic rather than Cretaceous in age (Fowell et al., 2011).

In the mid-1980s, Henry Roehler and Gary Stricker reported on dinosaurian fossil skin impressions and footprints from the Cretaceous of northwestern Alaska (Roehler and Stricker, 1984). Despite not being described in detail, the figures provided by the authors suggest that these remains can be attributed to ornithopods (Figure 2.6).

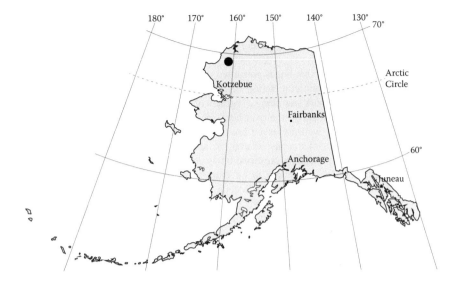

**FIGURE 2.6** Map showing the general location of dinosaur skin impressions and footprints reported by Roehler and Stricker (1984).

At approximately the same time, Louie Marincovich, Jr., of the USGS, brought the bones found by Robert Liscomb to the attention of Charles Repenning (Figure 2.7), also of the USGS, who properly identified the bones as dinosaurian. Louie Marincovich graciously granted permission for me to quote directly from his 1997 unpublished account of the first days of discovery regarding the correct identification of Liscomb's collection of fossil bones.

**FIGURE 2.7** Charles Repenning, the vertebrate paleontologist who recognized the first dinosaur bones from Alaska. The bones were originally identified by exploration geologist Robert Liscomb as mammoth bones. (Photograph courtesy of Louie Marincovich, Jr., United States Geological Survey [retired].)

Dinosaur bones were first collected in northern Alaska in 1961 but were not recognized as such until 1983. The history of how a routine find of presumed mammoth remains became the revolutionary find of the northernmost dinosaur bones is recorded here so it is not forgotten.

It is now well known that bones presumed to be mammoth remains were collected in 1961 by Liscomb while conducting field studies for Shell Oil Company (Davies, 1987). Part of the fieldwork was along a segment of the Colville River known as the Big Bend (owing to the S-shape of the river here), or as Ocean Point, a nearby geographic feature. Exceptionally well-preserved bones were collected at water level along the Colville River in beds of unconsolidated silt and sand at the base of 40-mile-high bluffs. Owing to their excellent preservation as unlithified remains in soft sediments, the bones evidently were thought to be mammoth remains in an especially thick section of the Gubik Formation, of Pliocene and Pleistocene age, or to have weathered out of the Gubik and fallen onto the Cretaceous sediments at river level. The Gubik Formation contains abundant bones of mammoths and other large mammals (Repenning, 1983) and mantles much of the northern Alaskan coastal plain, including the Ocean Point site.

The bones eventually went to storage in a warehouse in Houston, Texas where they remained half-forgotten until 1983. In October of that year, I gave a talk in Washington, DC on marine mollusks and microfossils from beds at Ocean Point that stratigraphically overlie the bone-bearing beds (Marincovich et al., 1983). Shortly after that, Richard V. Emmons, a paleontologist with Shell Oil Company in Houston, saw the abstract of this talk and called me to discuss the geology in the Ocean Point region. He asked in passing if I was interested in some mammoth bones from beds at Ocean Point. I agreed, with reluctance, to have the bones sent to me at the USGS office in Menlo Park, California, where I knew the resident vertebrate paleontologist, Charles Repenning, had already amassed a lifetime supply of mammoth remains.

Some weeks later, two large boxes of bones arrived in Menlo Park. I passed them on to Repenning, with apologies for foisting yet more mammoth bones on him. A few days later, Repenning asked me to walk over to his office to see a "surprise." What Chuck Repenning had neatly laid out was an arrangement of bones that made a triangular three-toed foot. All smiles, he asked me to quickly call Dick Emmons and get hold of more remains of "three-toed mammoths" which, he said, appeared to be hadrosaurian dinosaurs. So it happened that Chuck Repenning, in November 1983, became the first to recognize the remains of dinosaur bones from northern Alaska.

Repenning learned from Chris McGowan of the Royal Ontario Museum, Toronto, who was compiling northern dinosaur records at the time, that no dinosaur bones had been found as far north as Ocean Point (Figure 2.8). Subsequently, Repenning contacted Wann Langston at the University of Texas, Austin, who had considerable experience with hadrosaurs, and sent the Alaskan fossils to him. Shell Oil in Houston generously donated the specimens to the University of Texas for further study. The results of this earliest study of the Ocean Point dinosaurs collected by Liscomb were published by Langston's student, Kyle Davies, in 1987.

In 1984, a USGS field party organized by me and led by Elisabeth Brouwers (USGS, Denver) returned to Ocean Point to relocate the dinosaur-bearing strata. Brouwers' search by helicopter and foot was guided by a topographic map and oblique aerial photos that Dick Emmons had marked to show the general location of the bone site along a mile-long stretch of river bank. She and her field assistants soon located float specimens of dinosaur bones on the river band and nearby sand bars. The beds from which the specimens had weathered were also located. This party collected a representative

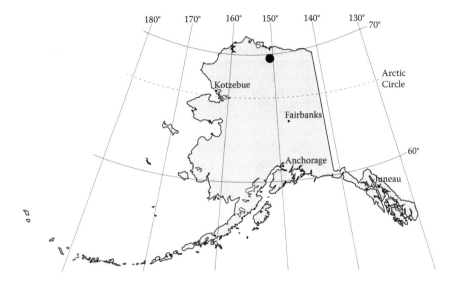

**FIGURE 2.8** Map showing the general location of the original discovery of bones by Robert Liscomb. The bones were later identified as belonging to duck-billed dinosaurs.

sample of dinosaur bones as well as a tooth of *Tyrannosaurus rex* (as later determined), and precisely recorded the location of the bonebeds for future seekers.

The USGS had, and has, no dinosaur specialist to pursue this study. Discovery of these far-northern dinosaurs had by then caused great excitement among the limited number of people who knew of it. No publicity was given to the find, in order to prevent vandalism of the site prior to scientific study. Seed money for a reconnaissance of the site in 1985 was provided by the USGS's Branch of Paleontology and Stratigraphy, and study of the Ocean Point bones was given over to William Clemens of the University of California at Berkeley. Owing to the rapid approach of the short field season in arctic Alaska, George Gryc, then chief of the Branch of Alaskan Geology, encouraged Clemens to quickly submit a proposal, and he expedited contractual arrangements with the USGS. A number of necessary official approvals that normally would have taken weeks or months to get were obtained by Gryc in days, by his walking the paperwork from one approving office to another. The 1985 reconnaissance (Clemens and Allison, 1985), made possible by one-time USGS funding, laid the foundation for extensive future collecting and study of the Ocean Point dinosaur site, which was thereafter largely funded by National Science Foundation grants.

Ensuing study of the Alaskan dinosaurs, other terrestrial and marine fossils, and their associated sediments, has made Ocean Point a world-class fossil site for both vertebrate and invertebrate paleontologists (Marincovich et al., 1990; Marincovich, 1993). This happy result sprang entirely from Chuck Repenning's recognition of "three-toed mammoths" as hadrosaurian dinosaurs, 22 years after they had been collected. His role in discovering these northernmost dinosaur bones should not be forgotten.

Thanks to Louis Taylor's coordinating contact information, late in 2004, Charles Repenning wrote to thank me for acknowledging his role in that early discovery (Fiorillo, 2004). He elaborated further, writing:

When I first opened the box (about $1 \times 1' \times 2'$ in size ...) of fossils that Shell sent me I was surprised. I had just published a report on Pliocene mammals from the Gubic Formation at nearby Ocean Point and expected to see more of the same. But they certainly weren't mammals and I quickly recognized the odd foot bones of a hadrosaur. The rest I didn't try to identify and assumed they belonged to the hadrosaur. I sent them all to Wann Langston because he had worked on hadrosaurs.

Those remains were formally described by Davies (1987) and attributed to hadrosaurs. Later study refined the taxonomy and the hadrosaur remains from the North Slope, that portion of Alaska between the Brooks Range and the Arctic Ocean, were attributed to the genus *Edmontosaurus* (Gangloff and Fiorillo, 2010; see Chapter 4). The year 1987 also saw additional publications on Cretaceous Alaskan vertebrates (Brouwers et al., 1987; Parrish et al., 1987), demonstrating that this region was likely to prove fruitful. In addition to these dinosaur remains, the field parties led by the University of California at Berkeley added fossil mammals—multituberculates, marsupials, and placentals—to this northern fauna (Clemens and Nelms, 1993).

Other institutions such as the University of Alaska Museum of the North and the Perot Museum of Nature and Science continued work on Cretaceous rocks of northern Alaska, including the now famous Liscomb Bonebed (Figure 2.9). Thus, most work has occurred along the bluffs of the Prince Creek Formation along the Colville River (Davies, 1987; Parrish et al., 1987; Clemens and Nelms, 1993; Gangloff, 1995, 1998; Fiorillo and Gangloff, 2000, 2001; Rich et al., 2002; Fiorillo, 2004, 2006, 2008; Gangloff et al., 2005; Fiorillo et al., 2009a,b, 2010a,b; Gangloff and Fiorillo, 2010; Brown and Druckenmiller, 2011; Fiorillo and Tykoski, 2012; 2013; 2014). In addition to the hadrosaur and mammal remains described thus far, the fauna recovered

**FIGURE 2.9**  A field crew from the Perot Museum of Nature and Science, Dallas, Texas, excavating dinosaur bones from what is now known as the Liscomb Bonebed.

from quarry excavations and accumulated river bar and bank float now includes specimens of chondrichthyan and osteichthyan fishes, large and small theropods, a hypsilophodontid, a pachycephalosaur, and hadrosaurian dinosaurs (Gangloff, 1998; Fiorillo and Gangloff, 2000, 2001; Gangloff et al., 2005; Gangloff and Fiorillo, 2010; Brown and Druckenmiller, 2011) as well as details of a horned dinosaur named in 2012 (Fiorillo and Tykoski, 2012, 2013; Tykoski and Fiorillo, 2013), and a tyrannosaur named in 2014 (Fiorillo and Tykoski, 2014).

These studies have established the existence of richer, more complex terrestrial Cretaceous biotic communities than had previously been recognized at this latitude. Moreover, additional taxa continue to be found (Gangloff et al., 2005; Fiorillo and Tykoski, 2012, 2014). However, even with all of the promise that this one area of northern Alaska holds along the Colville River, other areas of Alaska—including elsewhere in this northern region—also show great potential for dinosaur discoveries (Fiorillo, 2004, 2006; Fiorillo et al., 2010b).

That potential, in combination with the enduring public appeal of dinosaurs, inspired the National Park Service (NPS) Alaska Region to take a bold step and initiate paleontological surveys in Alaska's national parks. In 1999, Russell Kucinski, (Figure 2.10) then the team manager for Natural Resources in the NPS Alaska Region, invited NPS paleontologist Vincent Santucci and me to Anchorage to discuss the merits of undertaking such surveys. What I had expected to be a meeting of perhaps an hour stretched into a full day as enthusiasm for initiating these surveys built.

As a result of those discussions, we were invited back to Anchorage later that fall to attend a workshop and make a more formal presentation to a gathering of NPS

**FIGURE 2.10** Russell Kucinski, retired National Park Service, in Wrangell–St. Elias National Park, Alaska. Kucinski was an early advocate for investigating for dinosaurs within the Cretaceous rocks that occur in Alaskan National Parks. (Photograph courtesy of Linda Stromquist, National Park Service [retired].)

resource managers. Vince Santucci and I thought it appropriate to invite Roland Gangloff, then of the University of Alaska Museum of the North, to join us. The time slot given to us was on the afternoon of the third and final day of the workshop.

The group's reception to our presentation went as one could expect any presentation to be received at the end of the last day of a group cooped up in a small room for several days—or so it seemed. After the workshop broke up, however, two people approached me individually and introduced themselves: Peter Armato, who at the time was Chief of Resources at Kenai Fjords National Park, and Phil Brease, a geologist working in Denali National Park. Peter was intrigued by the possibilities of new fossil discoveries and he brainstormed with me about how to get such a survey funded. The parks in Alaska are arranged by networks, and Kenai Fjords National Park is part of the Southwest Network which also includes Lake Clark National Park, Katmai National Park, and Aniakchak National Monument (Figure 2.5). He proposed a strategy by which funding might be available if approached at a network level. For his part, Phil proposed to fund a small survey in Denali through some discretionary funding. As it turned out, both strategies succeeded and the surveys yielded significant and often spectacular results.

Thanks to initial funding through Russ Kucinski, in 2000 Vince Santucci and I undertook a preliminary reconnaissance tour of the Southwest Network parks with the purpose of introducing ourselves and our intentions if future funding developed. During the summer of 2001, a more formal effort began with a joint Perot Museum of Nature and Science and NPS Alaska Region paleontological survey of Aniakchak National Monument. The first several days of rafting the Aniakchak River, though exhilarating, yielded minimal paleontological results, certainly none to generate the excitement needed to continue investigating. On the final day, however, within 2 hours of our scheduled float plane pickup, one last effort to look around one more corner of a cliff face produced the first evidence of dinosaurs in a national park unit in the state: a series of three-toed footprints attributable to hadrosaurs (Fiorillo and Parrish, 2004). These tracks are in the Chignik Formation, a rock unit considered to have been deposited approximately 70 million years ago, making this sequence of rocks similar in age and depositional setting to the Prince Creek Formation on Alaska's North Slope. Without this discovery, it is difficult now to imagine having been able to sustain that initial interest in searching for dinosaurs in Alaska's national park units, because this discovery sustained and ignited new interest in expanding the search.

Subsequent paleontological surveys in latest Cretaceous rocks within Denali National Park in the central Alaska Range have produced a wealth of information regarding biodiversity of fossil vertebrates and invertebrates (Fiorillo et al., 2007, 2009b, 2011; Fiorillo and Adams, 2012; Fiorillo et al., 2014a,b, 2015, 2016; Fiorillo and Tykoski, 2016) as well as plants (Tomsich et al., 2010, 2014). The first-recognized dinosaur track discovery in the park was made during a geology field trip led by Paul McCarthy, a geology professor at the University of Alaska in 2005. As he was lecturing at an outcrop, describing the rocks and how they held potential for dinosaur footprints, two students, Carla Susi Tomsich and Jeremiah Drewel, pointed to a small three-toed track impression. Phil Brease alerted me to news of that track prior to my arrival in the park later that summer. Unable to anticipate whether additional tracks— let alone trackways—would be found in the park, Phil arranged for us to excavate that footprint when we arrived with the intention that it would eventually be on display in the Murie Science and Learning Center at the visitor center in Denali Park.

The Cretaceous rocks of Denali National Park contain one of the best records anywhere in the world from an ancient terrestrial ecosystem. Although the record here consists mostly of footprints, trackways, and other traces, these rocks provide valuable insights, as they record the presence of fossil vertebrates such as pterosaurs, birds, and fishes that have not been found anywhere else in the region (Fiorillo et al., 2009b, 2011; Fiorillo and Adams, 2012; Fiorillo et al., 2015). These finds perfectly complement the discoveries of time-equivalent bones from northern Alaska. Further, the tracks have amply demonstrated the presence of a remarkably rich avian fauna that is perhaps one of the best known from a single rock formation anywhere in the world (Fiorillo et al., 2011).

Similarly, in Wrangell–St. Elias National Park and Preserve, another joint Perot Museum of Nature and Science and NPS Alaska Region survey has yielded the footprints of small theropods and ornithopods in a Cretaceous rock unit (Fiorillo et al., 2012; Figure 2.11). These particular rocks in this park are so poorly understood that this rock unit has not been formally named, and precise chronostratigraphy is still unavailable. Structural correlations, however, suggest this unit is latest Cretaceous in age, tentatively placed between 75 and 67 million years ago (Trop and Ridgway, 2007). Collectively, the additional discoveries in other parks, combined with the initial finds along the Colville River, demonstrate the impressive extent of latest Cretaceous high-latitude terrestrial biotic communities in this region.

In addition to these spectacular and scientifically rich regions within Alaska, isolated Cretaceous dinosaurs have been found in a handful of other localities around the state. For example, a skull of the armored dinosaur *Edmontonia* was found weathering out of a concretion within the Matanuska Formation by hunting

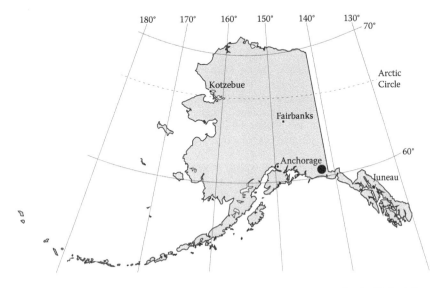

**FIGURE 2.11** Map showing the location of the largest national park in the United States, Wrangell–St. Elias National Park and Preserve.

guides John A. and John Joe Luster from Wasilla, Alaska. Their discovery is thought to have come from the Matanuska Formation of south-central Alaska (Gangloff, 1995), a rock unit that spans much of the Campanian and Maastrichtian stages of the Cretaceous, or from approximately 83 million years ago to approximately 67 million years ago (Trop and Ridgway, 2007). This ankylosaur specimen is the only documented occurrence of not only this animal in the region (Gangloff, 1995) but also the broader group of ankylosaurs.

Also of Cretaceous age was the discovery in the Talkeetna Mountains of a dinosaur skeleton, tentatively assigned to the Hadrosauria (Pasch and May, 1997). This dinosaur is from rocks thought to be Turonian in age, or approximately 94 to 89 million years in old. The specimen has yet to be fully described, but given its latitudinal position in the Cretaceous, the specimen promises to be significant in understanding the biogeography of hadrosaurs in North America, including the geologically younger hadrosaurs found at the site of Liscomb's original discovery.

All of these discoveries have generated a keen interest in Alaska. It is nevertheless worth emphasizing that these discoveries are part of a continuum of discovery ultimately connected to a long-standing fascination with the Far North. From the earliest explorations, scientific discovery in the Arctic has perplexed and exhilarated scientists as well as the public in general. Carl Benson's statement that "ignorance of the Arctic is an infinite resource" (Norton, 2011), suggests that such future discoveries will likely continue to perplex and exhilarate.

Alaska contains 20% of all of the terrestrial surface area of the United States. With such expansive geography, the state promises to be the source of more new discoveries furthering the continuum, and reminiscent in some ways of the richness experienced by the first paleontological explorers of the U.S. Rocky Mountain west. Each new discovery helps us further refine our understanding of this unique ancient high-latitude terrestrial ecosystem and offers insights into how these biotic communities developed.

## REFERENCES

Althoff, W. 2007. *Drift Station: Arctic Outposts of Superpower Science.* Dulles, VA: University of Potomac Books.

Beechey, F.W. 1831. *Narrative of a Voyage to the Pacific and Beering's Strait.* London, UK: Henry Colburn and Richard Bentley, 2 volumes.

Benton, M.J. 2004. Origin and relationships of Dinosauria. In *The Dinosauria*, eds. D.B. Weishampel, P. Dodson, and H. Osmolska, 7–19. Berkeley, CA: University of California Press.

Brouwers, E.M., W.A. Clemens, R.A. Spicer, T.A. Ager, L.D. Carter, and W.V. Sliter. 1987. Dinosaurs on the North Slope, Alaska: High latitude, latest Cretaceous environments. *Science* 237:1608–1610.

Brown, C.M. and P. Druckenmiller. 2011. Basal ornithopod (Dinosauria: Ornithischia) teeth from the Prince Creek Formation (early Maastrichtian) of Alaska. *Canadian Journal of Earth Sciences* 48:1342–1354.

Buckland, W. 1824. Notice on the Megalosaurus or great fossil lizard of Stonesfield. *Transactions of the Geological Society*, second series 1:390–396.

Campbell, L.J. 1994. The terrible lizards. *Alaska Geographic* 21(4):24–37.

Chevigny, H. 1965. *Russian America: The Great Alaskan Venture, 1741–1867.* New York, NY: Viking Press.

Clemens, W.A. 1994. Continental vertebrates from the Late Cretaceous of the North Slope, Alaska. *1992. Proceedings International Conference on Arctic Margins*, eds. D.K. Thurston and K. Fujita, 395–398. Outer Continental Shelf Study, Mineral Management Service, 94–0040. Anchorage, AK.

Clemens, W.A. and C.W. Allison. 1985. Late Cretaceous terrestrial vertebrate fauna, North Slope, Alaska. *Geological Society of America Abstracts with Programs* 17:548.

Clemens, W.A. and L.G. Nelms. 1993. Paleoecological implications of Alaskan terrestrial vertebrate fauna in latest Cretaceous time at high paleolatitudes. *Geology* 21:503–506.

Colbert, E.H. 1961. *Dinosaurs: Their Discovery and Their World.* New York, NY: E.P. Dutton & Co., Inc.

Davies, K.L. 1987. Duck-billed dinosaurs (Hadrosauridae: Ornithischia) from the North Slope of Alaska. *Journal of Paleontology* 61:198–200.

Dodson, P. 1990. Counting dinosaurs: How many kinds were there? *Proceedings of the National Academy of Sciences* 87:7608–7612.

Evans, D.C., M.J. Vavrek, D.R. Braman, N.E. Campione, T.A. Dececchi, and G.D. Zazula. 2012. Vertebrate fossils (Dinosauria) from the Bonnet Plume Formation, Yukon Territory, Canada. *Canadian Journal of Earth Sciences* 49:396–411.

Fiorillo, A.R. 2004. The dinosaurs of Arctic Alaska. *Scientific American* 291:84–91.

Fiorillo, A.R. 2006. Review of the Dinosaur Record of Alaska with comments regarding Korean dinosaurs as comparable high-latitude fossil faunas. *Journal of Paleontological Society of Korea* 22:15–27.

Fiorillo, A.R. 2008. On the occurrence of exceptionally large teeth of Troodon (Dinosauria:Saurischia) from the Late Cretaceous of northern Alaska. *Palaios* 23:322–328.

Fiorillo, A.R. and T.L. Adams. 2012. A therizinosaur track from the Lower Cantwell Formation (Upper Cretaceous) of Denali National Park, Alaska. *Palaios* 27:395–400.

Fiorillo, A.R., T.L. Adams, and Y. Kobayashi. 2012. New sedimentological, palaeobotanical, and dinosaur ichnological data on the palaeoecology of an unnamed Late Cretaceous rock unit in Wrangell–St. Elias National Park and Preserve, Alaska, USA. *Cretaceous Research* 37:291–299.

Fiorillo, A.R., M. Contessi, Y. Kobayashi, and P.J. McCarthy. 2014a. Theropod tracks from the Lower Cantwell Formation (Upper Cretaceous) of Denali National Park, Alaska, USA with comments on theropod diversity in an ancient, high-latitude terrestrial ecosystem. In *Fossil Footprints of Western North America*, eds. M. Lockley and S.G. Lucas, 429–439. *New Mexico Museum of Natural History and Science Bulletin* 62. Albuquerque, NM.

Fiorillo, A.R., P.L. Decker, D.L. LePain, M. Wartes, and P.J. McCarthy. 2010a. A probable Neoceratopsian Manus Track from the Nanushuk Formation (Albian, Northern Alaska). *Journal of Iberian Geology* 36:165–174.

Fiorillo, A.R. and R.A. Gangloff. 2000. Theropod teeth from the Prince Creek Formation (Cretaceous) of northern Alaska, with speculations on arctic dinosaur paleoecology. *Journal of Vertebrate Paleontology* 20:675–682.

Fiorillo, A.R. and R.A. Gangloff. 2001. The caribou migration model for Arctic hadrosaurs (Ornithischia: Dinosauria): A reassessment. *Historical Biology* 15:323–334.

Fiorillo, A.R., S.T. Hasiotis, and Y. Kobayashi. 2014b. Herd structure in Late Cretaceous polar dinosaurs: A remarkable new dinosaur tracksite, Denali National Park, Alaska, USA. *Geology* 42:719–722.

Fiorillo, A.R., S.T. Hasiotis, Y. Kobayashi, and C.S. Tomsich. 2009a. A pterosaur manus track from Denali National Park, Alaska Range, Alaska, USA. *Palaios* 24:466–472.

Fiorillo, A.R., S.T. Hasiotis, Y. Kobayashi, B.H. Breithaupt, and P.J. McCarthy. 2011. Bird tracks for the Upper Cretaceous Cantwell Formation of Denali National Park, Alaska, USA: A new perspective on ancient polar vertebrate biodiversity. *Journal of Systematic Palaeontology* 9:33–49.

Fiorillo, A.R., Y. Kobayashi, P.J. McCarthy, T.C. Wright, and C.S. Tomsich. 2015. Reports of pterosaur tracks from the Lower Cantwell Formation (Campanian-Maastrichtian) of Denali National Park, Alaska, USA, with comments about landscape heterogeneity and habitat preference. *Historical Biology* 27:672–683.

Fiorillo, A.R., P.J. McCarthy, and S.T. Hasiotis. 2016. Crayfish burrows from the latest Cretaceous lower Cantwell Formation (Denali National Park, Alaska): Their morphology and paleoclimatic significance. *Palaeogeography, Palaeoclimatology, Palaeoecology* 441:352–359.

Fiorillo, A.R., P.J. McCarthy, B. Breithaupt, and P. Brease. 2007. Dinosauria and fossil Aves footprints from the Lower Cantwell Formation (latest Cretaceous), Denali Park and Preserve, Alaska. *Alaska Park Science* 6:41–43.

Fiorillo, A.R., P.J. McCarthy, P.P. Flaig, E. Brandlen, D.W. Norton, P. Zippi, L. Jacobs, and R.A. Gangloff. 2010b. Paleontology and paleoenvironmental interpretation of the Kikak-Tegoseak Quarry (Prince Creek Formation: Late Cretaceous), northern Alaska: A multi-disciplinary study of a high-latitude ceratopsian dinosaur bonebed. In *New Perspectives on Horned Dinosaurs*, eds. M.J. Ryan, B.J. Chinnery-Allgeier, and D.A. Eberth, 456–477. Bloomington, IN: Indiana University Press.

Fiorillo, A.R. and J.T. Parrish. 2004. The first record of a Cretaceous dinosaur from western Alaska. *Cretaceous Research* 25:453–458.

Fiorillo, A.R. and R.S. Tykoski. 2012. A new species of centrosaurine ceratopsid Pachyrhinosaurus from the North Slope (Prince Creek Formation: Maastrichtian) of Alaska. *Acta Palaeontologica Polonica* 57:561–573.

Fiorillo, A.R. and R.S. Tykoski. 2013. An immature *Pachyrhinosaurus perotorum* (Dinosauria: Ceratopsidae) nasal reveals unexpected complexity of craniofacial ontogeny and integument of Pachyrhinosaurus. *PLoS ONE* 8(6):e65802. doi:10.1371/journal.pone.0065802.

Fiorillo, A.R. and R.S. Tykoski. 2014. A diminutive new tyrannosaur from the top of the world. *PLoS ONE* 9(3):e91287. doi:10.1371/journal.pone.0091287.

Fiorillo, A.R. and R.S. Tykoski. 2016. Small hadrosaur manus and pes tracks from the lower Cantwell Formation (Upper Cretaceous), Denali National Park, Alaska: Implications for locomotion in juvenile hadrosaurs. *Palaios* 31:479–482.

Fiorillo, A.R., R.S. Tykoski, P.J. Currie, P.J. McCarthy, and P. Flaig. 2009b. Description of two partial troodon braincases from the Prince Creek Formation (Upper Cretaceous), North Slope Alaska. *Journal of Vertebrate Paleontology* 29:178–187.

Ford, C. 1966. *Where the Sea Breaks Its Back: The Epic Story of a Pioneer Naturalist and the Discovery of Alaska.* Boston, MA: Little, Brown.

Fowell, S.J., P. Druckenmiller, P.J. McCarthy, R.B. Blodgett, and K. May. 2011. Paleoecology of Alaska's Jurassic Park. *Geological Society of America Abstracts with Programs* 43:264.

Frolov, I.E., Z.M. Gudkovich, V.F. Radionov, A.V. Shirochkov, and L.A. Timokhov. 2005. *The Arctic Basin Results from the Russian Drifting Stations.* Berlin, Germany: Springer-Verlag, and Chichester, UK: Praxis Publishing.

Gangloff, R.A. 1995. Edmontonia sp., the first record of an ankylosaur from Alaska. *Journal of Vertebrate Paleontology* 15:195–200.

Gangloff, R.A. 1998. Arctic dinosaurs with emphasis on the Cretaceous record of Alaska and the Eurasian-North American connection. In *Lower and Middle Cretaceous Terrestrial Ecosystems*, eds. S.G. Lucas, J.I. Kirkland, and J.W. Estep, 211–220. New Mexico Museum of Natural History and Science Bulletin No. 14. Albuquerque, NM.

Gangloff, R.A. 2012. Dinosaurs under the Aurora. Bloomington, IN: Indiana University Press.

Gangloff, R.A. and A.R. Fiorillo. 2010. Taphonomy and paleoecology of a bonebed from the Prince Creek Formation, North Slope, Alaska. *Palaios* 25:299–317.

Gangloff, R.A., A.R. Fiorillo, and D.W. Norton. 2005. The first Pachycephalosaurine (Dinosauria) from the Paleo-Arctic and its paleogeographic implications. *Journal of Paleontology* 79:997–1001.

Heintz, N. 1963. Dinosaur footprints and polar wandering. *Norsk Polarinstitutt Årbok* 1962:35–43.

Hoffecker, J.F., W.R. Powers, and T. Goebel. 1993. The colonization of Beringia and the peopling of the New World. *Science* 259:46–43.

Holmes, T. and P. Dodson. 1997. Counting more dinosaurs—How many kinds are there. In *Dinofest International: Proceedings of a Symposium Sponsored by Arizona State University*, eds. D.L. Wolberg, E. Stump, and G.D. Rosenberg, 125–128. Philadelphia, PA: Academy of Natural Sciences.

Kotzebue, O. von 1821. *A Voyage of Discovery into the South Sea and Beering's Straits.* London, UK: Longman, Hurst, Rees, Orme, and Brown, Paternoster-Row, 3 volumes.

Lapparent, A.F. de. 1962. Footprints of Dinosaur in the Lower Cretaceous of Vestspitsbergen-Svalbard. *Norsk Polarinstitutt Årbok* 1960:14–21.

Leidy, J. 1856. Notices of remains of extinct reptiles and fishes, discovered by Dr. F.V. Hayden in the Bad Lands of the Judith River, Nebraska Territory. *Proceedings of the Academy of Natural Sciences of Philadelphia* 8:72–73.

Mantell, G.A. 1825. Notice on the Iguanodon, a newly discovered fossil reptile, from the sandstone of the Tilgate Forest, in Sussex. *Philosophical Transactions of the Royal Society of London* 115:179–186.

Marincovich, L., Jr. 1993. Danian mollusks from the Prince Creek Formation, northern Alaska, and implications for Arctic Ocean paleogeography. *Paleontology Society Memoir* 35:1–35.

Marincovich, L., Jr., E.M. Brouwers, and D.M. Hopkins. 1983. Paleogeographic affinities and endemism of Cretaceous and Paleocene marine fauns in the Arctic. *United States Geological Survey Circular* 911:45–46.

Marincovich, L., Jr., E.M. Brouwers, D.M. Hopkins, and M.C. McKenna. 1990. Late Mesozoic and Cenozoic paleogeographic and paleoclimatic history of the Arctic Ocean Basin, based on shallow-water marine faunas and terrestrial vertebrates. In *The Arctic Ocean Region*, eds. A. Grantz, L. Johnson, and J.F. Sweeney, 403–426. Boulder, CO: Geological Society of America, The Geology of North America, vol. L.

McClellan, P.H. 1980. Does the Alaskan record of land vertebrate fossils substantiate or contradict the long postulated existence of a Bering Land Bridge before the Pleistocene? *Alaskan Journal of Pre-Pleistocene Vertebrate Paleontology: "The Thinnest Journal in the World"* 1:4–17. Self-published for the Western Association of Vertebrate Paleontology.

Mirsky, J. 1948. *To the North! The Story of Arctic Exploration from Earliest Times to the Present.* Chicago, IL: The University of Chicago Press.

Norton, D.W. 2001. *Fifty More Years Below Zero: Tributes and Meditations for the Naval Arctic Research Laboratory's First Half Century at Barrow, Alaska.* Calgary, AB: The Arctic Institute of North America.

Norton, D.W. 2011. Review of Siku: Knowing our ice. Documenting Inuit Sea-Ice Knowledge and Use, eds. I. Krupnik, C. Aporta, S. Gearheard, G.J. Laidler, and L.K. Holm, 501pp. Dordrecht, Germany: Springer, 2010. Arctic 64:381–384.

Owen, R. 1842. Report on British fossil reptiles, part II. *Report of the Eleventh Meeting of the British Association for the Advancement of Science.* London, UK: John Murray. 60–204.

Parrish, M.J., J.T. Parrish, J.H. Hutchinson, and R.A. Spicer. 1987. Late Cretaceous vertebrate fossils from the North Slope of Alaska and implications for dinosaur ecology. *Palaios* 2:377–389.

Pasch, A.D. and K.C. May. 1997. First occurrence of a hadrosaur (Dinosauria) from the Matanuska Formation (Turonian) in the Talkeetna Mountains of south-central Alaska. In *Short Notes on Alaska Geology, 1997*, eds. J.G. Clough and F. Larson, 99–109. Fairbanks, AK: Alaska Department of Natural Resources, Professional Report 118.

Pitulko, V.V., P.A. Nikolsky, E. Yu. Gira, A.E. Basilyan, V.E. Tumskoy, S.A. Koulakov, S.N. Astakhov, E. Yu. Pavlova, and M.A. Anisimov. 2004. The Yana RHS Site: Humans in the Arctic before the last glacial maximum. *Science* 303:52–56.

Preston, D.J. 1986. *Dinosaurs in the Attic: An Excursion into the American Museum of Natural History*. New York, NY: St. Martin's Press.

Rainger, R. 1991. *An Agenda for Antiquity: Henry Fairfield Osborn and Vertebrate Paleontology at the American Museum of Natural History, 1890–1935*. Tuscaloosa, AL: University of Alabama Press.

Repenning, C.A. 1983. New evidence for the age of the Gubik Formation, Alaskan North Slope. *Quaternary Research* 19:356–372.

Rich, T.H., P. Vickers-Rich, and R.A. Gangloff. 2002. Polar dinosaurs. *Science* 295:979–980.

Roderick, J. 1997. *Crude Dreams: A Personal History of Oil & Politics in Alaska*. Kenmore, WA: Epicenter Press, Inc.

Roehler, H.W. and G.D. Stricker. 1984. Dinosaur and wood fossils from the Cretaceous Corwin Formation in the National Petroleum Reserve, North Slope, Alaska. *Journal of the Alaska Geological Society* 4:35–41.

Rouse, G.E. and S.K. Svrivastava. 1972. Palynological zonation of Cretaceous and early Tertiary rocks of the Bonnet Plume Formation Northeastern Yukon, Canada. *Canadian Journal of Earth Sciences* 9:1163–1179.

Thomson, K. 2008. *The Legacy of the Mastodon: The Golden Age of Fossils in North America*. New Haven: Yale University Press.

Tomsich, C.S., P.J. McCarthy, S.J. Fowell, and D. Sunderlin. 2010. Paleofloristic and paleoenvironmental information from a Late Cretaceous (Maastrichtian) flora of the lower Cantwell Formation near Sable Mountain, Denali National Park, Alaska. *Palaeogeography, Palaeoclimatology, Palaeoecology* 295:389–408.

Tomsich, C.S., P.J. McCarthy, A.R. Fiorillo, D.B. Stone, J.A. Benowitz, and P.B. O'Sullivan. 2014. New zircon U-Pb ages for the lower Cantwell Formation: Implications for the Late Cretaceous paleoecology and paleoenvironment of the lower Cantwell Formation near Sable Mountain, Denali National Park and Preserve, central Alaska Range, USA. In *Proceedings of the International Conference on Arctic Margins VI*, Fairbanks, Alaska, May 2011, eds. D.B. Stone, G.K. Grikurov, J.G. Clough, G.N. Oakey, and D.K. Thurston, 19–60. St. Petersburg, Russia: VSEGEI.

Trop, J.M. and K.D. Ridgway. 2007. Mesozoic and Cenozoic tectonic growth of southern Alaska: A sedimentary basin perspective. In *Tectonic Growth of a Collisional Continental Margin: Crustal Evolution of Southern Alaska*, eds. K.D. Ridgway, J.M. Trop, J.M.G. Glen, and J.M. O'Neill, 55–94. Boulder, CO: Geological Society of America Special Paper 431.

Tykoski, R.S. and A.R. Fiorillo. 2013. The braincase of *Pachyrhinosaurus perotorum* compared to other *Pachyrhinosaurus* species, and its utility for species-level recognition. *Earth and Environmental Science Transactions of the Royal Society of Edinburgh* 103:487–499.

Wang, S.C. and P. Dodson. 2006. Estimating the diversity of dinosaurs. *Proceedings of the National Academy of Sciences* 103:13601–13605.

Weishampel, D.B., P.M. Barrett, R.E. Coria, J. Le Loeuff, E.S. Gomani, Z. Zhao, X. Xu, A. Sahni, and C. Noto. 2004. Dinosaur distribution. In *The Dinosauria*. 2nd edition, eds. D.B. Weishampel, P. Dodson, and H. Osmólska, 517–606. Berkeley, CA: University of California Press.

# 3 A Paleontologist's Perspective on the Geology of Alaska

Dinosaurs first appeared approximately 230 million years ago (Rogers et al., 1993; Martinez et al., 2011) in the Triassic Period, though some have suggested that the first appearance of dinosaurs may have occurred even earlier (Nesbitt et al., 2012). Dinosaurs then ruled as the dominant vertebrate life forms for the next 160 million years, and now exist as birds. Dinosaurs have been found on all continents from Mesozoic-aged rocks, and all of these discoveries have contributed to our understanding of this remarkable group of animals. During the Mesozoic window of time the distribution of land that now comprises Alaska changed dramatically, and it is that change that makes the Cretaceous Alaskan dinosaur story important. As we shall see here, the Cretaceous dinosaur record of Alaska is unique in its paleogeography and this has significant ramifications as to what these high-latitude dinosaurs tell us. No place else on the planet Earth records such a complete record of a high-latitude dinosaur ecosystem.

Shortly after the concept of plate tectonics was developed, geologists recognized that the Cordilleran province of western North America was a complex system of geologic blocks, or terranes (e.g., Dewey and Bird, 1970; Coney et al., 1980, 1994; Burchfiel et al., 1992). These discrete, fault-bounded terranes have been likened to pieces of a puzzle comprising the western edge of the continent (Blodgett and Stanley, 2008). The accepted model for North America was that the western part of the continent was built by the convergence from the west of relatively small continental plates colliding with the accreting continental edge, and at times some of these relatively smaller plates had previously collided to form larger plates, sometimes referred to as composite terranes, which in turn then collided with North America (e.g., Coney et al., 1980; Plafker and Berg, 1994a,b; Winkler, 2000). These collisions are often the result of the subduction of one lithospheric plate beneath another plate, drawing lighter continental crust together as oceanic crust is consumed. The result of this process not only brings together two disparate geologic provinces, but can also create basins of sedimentary deposition within the subduction zones. In addition to this process of collision, subduction of the oceanic plates beneath the continental plates creating huge bodies of intrusive magma, or batholiths, is observed throughout the Cordillera. Perhaps the most famous is the Sierra Nevada Batholith, which is so prominently exposed now in Yosemite National Park in California. Alaska represents the northern end of this enormous Cordilleran province, which extends to the southern tip of South America.

Alaska comprises approximately 20% of the geographic area of the United States, and as one would expect, the geologic history of the region, ranging from its tectonic history to its volcanic history, to its glacial history, is complex. Given that this work is focused on Cretaceous dinosaurs, their environments, and the climate in which they lived, it is beyond the scope of this work to provide a detailed summary of this complex regional geologic history, particularly when so many others already exist. So, for those keenly interested in details of Alaskan geology there is an array of overviews such as the statewide tectonic perspective by Plafker and Berg (1994a,b), the history of the Arctic margin by Miller et al. (2002), and the tectonic evolution of southern Alaska by Ridgway et al. (2007) or Winkler (2000).

Similarly, it is also worth noting that there is a rich, well-known, and diverse record of fossil vertebrates from the modern Arctic, particularly from units of Pliocene and Pleistocene age, commonly known as the Ice Age (e.g., Skinner and Kaisen, 1947; Guthrie and Matthews, 1971; Guthrie, 1990, 2003; Zazula et al., 2007; Rivals et al., 2010), but also from early to mid-Paleozoic age, approximately 433–360 million years ago (e.g., Stensiö, 1927; Ørvig, 1957, 1975; Denison, 1963; Broad and Lenz, 1972; Dineley and Loeffler, 1976; Elliott, 1983, 1984; Elliott and Dineley, 1983; Märss et al., 1998; Daeschler et al., 2006). Much of this record is from either north central or northeastern Canada, Greenland, or Spitzbergen (Norway), with some of these records within the North American Cordilleran province. Though fewer in number, there are also important records of vertebrate remains from places such as Yukon Territory in northwestern Canada (e.g., Denison, 1963; Broad and Lenz, 1972), Belarus, Estonia, Latvia, and Russia (e.g., Vorobyeva, 1980; Mark-Kurik, 2000; Blieck et al., 2002) that help elucidate the biodiversity and evolutionary relationships of vertebrates of the time. In contrast to these and many other well-documented records of Paleozoic early vertebrates in the Arctic, despite an abundance of rock of this age, Alaska has produced only very fragmentary fossil material of comparable age (Figure 3.1). These materials have a much less robust published

**FIGURE 3.1**   Paleozoic outcrops in Yukon–Charley Rivers National Preserve, east-central Alaska.

record than comparably aged fossil vertebrates found elsewhere in the Arctic (Brabb and Churkin, 1969; Perkins, 1971), and a few unpublished reports of the United States Geological Survey during early days of geologic investigation of the region (e.g., Dunkle, 1964). While important to the discussion regarding the evolution of life on Earth, the Paleozoic sedimentary sequences that have produced these fossils were deposited in equatorial to subequatorial environments, much farther to the south than where these sequences are found today, and tectonically drifted north since the Paleozoic (Lawver et al., 2002). Therefore, rather than attempt a detailed summary of all aspects of Alaskan geologic history, this chapter instead focuses on the geologic history relevant to this story about Alaskan dinosaurs.

Like the rest of the western edge of North America, Alaska is an amalgamation of puzzle pieces, and within some of these puzzle pieces, evidence for long distance transport can be readily found. Within Alaska some of the composite tectonic terranes contain smaller basins that were deposited near their present latitudinal position, but paleontological evidence has shown in some cases that the basement rock of the terranes originated thousands of kilometers away from its current position.

Many of the rocks exposed, for example, near the most populous part of Alaska, Anchorage, and seen by countless tourists as they rush to see wildlife or glaciers, belong to a geologic unit known as the McHugh Complex, named for rocks exposed along McHugh Creek south of town (Clark, 1973). These rocks (Figure 3.2) are part of the larger Chugach Terrane and represent a vast Mesozoic-aged subduction complex consisting largely of metamorphosed siltstones, sandstones, and conglomerates as well as some metamorphosed volcanic rocks. The McHugh Complex seems to have been deposited at two different geologic intervals, during the later Jurassic

**FIGURE 3.2**   McHugh Complex exposed along Exit Glacier in Kenai Fjords National Park. The McHugh Complex is a Mesozoic-aged sequence of metamorphosed siltstones, sandstones, conglomerates, and some metavolcanic rocks that were deposited as the result of subduction.

and in the Cretaceous (Amato and Pavlis, 2010). However, contained within this thick Mesozoic sequence of rocks are older house-sized to basketball-sized blocks of light-to-medium gray Permian limestone (Figure 3.3). Several of these blocks contain the one-celled marine organisms called foraminifera and the now extinct and unusual fusilinids, as well as alga and sponge-like animals, and conodonts, which

**FIGURE 3.3**  A small block of gray limestone within the McHugh Complex. These blocks range in size from small to house-sized and contain a number of marine fossils that suggest deposition under much warmer conditions. The fossils contained within these blocks have similarities to fossils found in central Asia, suggesting that the limestones originated far from their present location.

are primitive chordates, an early offshoot of the group to which we belong. This biota suggests a much warmer setting during deposition (Stevens et al., 1997; Fiorillo et al., 2004). The fusilinids in particular have a Tethyan affinity with faunal similarity to Permian assemblages in central Asia (Stevens et al., 1997), suggesting that these limestones very likely originated quite some distance from their present location.

Moving forward in geologic time to the Triassic Period, if one could view a satellite photo of Alaska some 225 million years ago the Alaskan landscape would not be recognizable, as much of the geographic area of modern Alaska was incomplete and missing some of the tectonic blocks that would arrive later. Although an abundance of Triassic rocks is present in the state, there are no dinosaurs of that age reported and vertebrate remains from those rocks as currently known are relatively uncommon. The few reports on Alaska Triassic vertebrates have typically been in reference to marine ichthyosaurs (e.g., Tailleur, 1973; Callaway and Massare, 1989; Druckenmiller et al., 2014), though recent work has shown great promise for improving the vertebrate record contained within the Triassic rocks of the region. Adams (2009) studied a Late Triassic vertebrate fauna from Hound Island in southeastern Alaska comprising various chondrichthyan and osteichtyan fishes and marine reptiles, including ichthyosaurs (Figure 3.4) and lesser known thalattosaurs. Within this fauna there is evidence, albeit very fragmentary, of the more derived Eusauropterygia, the marine reptile group that includes plesiosaurs.

This exciting fauna is from sediments included within the Hound Island Volcanics (Figure 3.5), which is part of the Alexander Terrane. The Alexander Terrane encompasses much of eastern and particularly southeastern Alaska, and illustrates the assembly of the southern margin of this part of North America. Paleomagnetic data from the Hound Island Volcanics place the Alexander Terrane at approximately

**FIGURE 3.4** Example of ichthyosaurs, marine reptiles that once inhabited the marine waters that covered parts of Alaska at one time. (Artwork by Karen Carr.)

**FIGURE 3.5**  Exposure of Triassic rocks found on Hound Island in southeastern Alaska. (Photograph courtesy of Thomas Adams, Witte Museum.)

latitude 10–20° north at the time of deposition, whereas the current latitude is approximately 56°, suggesting a northward drift of about 6000 kilometers, and deposition of the sediments that contain this Triassic vertebrate fauna were part of a subduction zone complex. More specifically, the deposition of these sediments occurred within an intra-arc basin, a basin that was the result of the back-arc rifting that occurred along the Alexander Terrane (Nokleberg et al., 2005) as it drew closer to the mainland of proto-Alaskan North America.

The vertebrate remains can be found in two different rock types, volcanic-rich limestones and calcareous shales, and because these two rock types represent two different depositional settings, the vertebrate fossil remains represent two different modes of fossil accumulation (Adams, 2009). The first mode of preservation is that of disarticulated skeletal remains. The nature of the disarticulation suggests accumulation in a higher-energy setting above the wave base of the marine environment, with subsequent gravity-displaced deposition such as turbidites and debris flows (Adams, 2009). The second mode of vertebrate fossil accumulation is the association of several partial skeletons of the large ichthyosaur, *Shonisaurus*. Given the associated nature of these skeletons within the calcareous shales, it is likely these animals were buried in a much lower-energy depositional environment below storm wave base along the outer portion of the outer slope of this shoreline.

It was not until the Middle Jurassic to the Early Cretaceous (sometime between 174–100 million years ago) that the Alexander Terrane was accreted to the rest of North America (Plafker and Berg, 1994a,b; Nokleberg et al., 2000), so it is not until the middle Cretaceous (100 million years ago), if we could look once again from an imaginary satellite, that we would see southeast Alaska has a more familiar appearance. Therefore these Triassic fossil vertebrates of the Alexander Terrane were hijacked from elsewhere and brought to the region long after burial of the fossils. Similarly, there is only a scant fossil vertebrate record from the Middle Jurassic—specifically about 168–166 million years ago—and again it is marine reptiles (Druckenmiller and Maxwell, 2014), and this record also represents fossil remains that were brought in from elsewhere by subsequent tectonic movements.

The Upper Jurassic Naknek Formation, a rock unit deposited approximately 160–148 million years ago, is the most widespread Mesozoic rock unit on the Alaska Peninsula (Detterman et al., 1981, 1996; Wilson et al., 1999), and it contains the earliest record of dinosaur remains from Alaska. Spurr (1900) named the formation during the first comprehensive geological survey of the region, and subsequent workers have modified and then subdivided the Naknek Formation into thinner units or members (Martin, 1905; Detterman et al., 1981, 1996; Wilson et al., 1999). These members are, from oldest to youngest, the Chisik Conglomerate Member, the Northeast Creek Sandstone Member, the Snug Harbor Siltstone Member, the Indecision Creek Sandstone Member, and the Katolinat Conglomerate Member. Together they represent several thousands of meters of stratigraphic section. In general, these members show a depositional change from a dominantly terrestrial fluvial system to a moderately shallow to deep marine environment. Based on marine invertebrate fossils, the age of the Naknek Formation is generally considered as Oxfordian to Tithonian. Radiometric date for the basal boundary of the Oxfordian is 161 million years and the date for the upper boundary of the Tithonian is 145 million years—dates that define the interval represented by these rocks as the Late Jurassic (Gradstein et al., 2004).

The Alaskan dinosaur record for the Jurassic Period is very limited; there are only two known Jurassic dinosaur localities in the state. One locality, a tracksite, was documented by photographs in the popular literature (Conyers, 1978; Campbell, 1994) showing a series of medium-sized, three-toed theropod footprints. The other is a robust but fragmentary bone cobble, a specimen that has only been illustrated but not described in technical detail (Fiorillo et al., 2004). Both records are from the Upper Jurassic Naknek Formation in southwestern Alaska along the Alaska Peninsula (Figures 3.6 and 3.7), with the latter record coming from deposits of the Snug Harbor Siltstone Member in Katmai National Park that represent a prograding delta complex (Fiorillo, 2006). While a study of similar deposits of the Naknek Formation may prove fruitful in furthering our understanding of Jurassic dinosaurs from Alaska, these rocks were deposited some distance south of where they are currently located.

During the first part of the Cretaceous, Alaska was close to being assembled (Figure 3.8). The region was essentially caught between the Pacific Plate pushing Alaska to the north while an opening rift in the Arctic Basin was pushing northern Alaska to the south (Lawver and Scotese, 1990; Lawver et al., 2002). These forces combined in the Cretaceous to produce a landscape that existed in the ancient high latitudes, essentially at the latitudes where it is found now, and therefore the Early Cretaceous dinosaur record in Alaska is the oldest basis for understanding the paleobiology of ancient polar terrestrial residents.

Most records of earlier Cretaceous dinosaurs occur in the northern part of the state and are from the Nanushuk Formation (Figure 3.7), a rock unit consisting of a thick

**FIGURE 3.6** Exposures of rocks in the Valley of 10,000 Smokes in Katmai National Park and Preserve. The dark gray rocks in the foreground are the Upper Jurassic Naknek Formation. The lighter exposures in the background are deposits laid down as the result of the 1912 Novarupta volcanic eruption, the largest such eruption to have occurred in the 20th century.

series of coarse to fine-grained clastic rocks interbedded with coals. These rocks were deposited in a paleogeographic position perhaps as high as 85° N (Witte et al., 1987). Originally named as the Nanushuk series by Schrader (1902), the unit was elevated to group status by Gryc et al. (1951). Previous discussions of dinosaur sites in northern Alaska from this earlier window of the Cretaceous Period has used stratigraphic terms such as the Corwin Formation (Roehler and Stricker, 1984) and the Chandler Formation (Gangloff, 1994, 1998; Parrish et al., 1987), and work in the Corwin Formation is of historical significance because it is the first technical report of dinosaur remains from the state (see Chapter 2). Both the Corwin Formation and Chandler Formation designations have since been abandoned, along with the term Nanushuk Group, with the revision of stratigraphic nomenclature of Cretaceous and Tertiary rock units in northern Alaska by Mull and colleagues (2003). In their revision of the stratigraphy of the rocks of these ages in northern Alaska, the old Corwin and Chandler formations are now considered as within the Nanushuk Formation (Mull et al., 2003). Though the Cretaceous record of Alaska has been dominated by reports of footprints, including the Nanushuk Formation (e.g., Fiorillo et al., 2010a), there are also reports of fragmentary dinosaur bones from this rock unit (Parrish et al., 1987; Fiorillo, 2006), suggesting that a more detailed paleontological story still waits to be told.

The modern Arctic is not a homogenous environment but rather a mix of environments within both the terrestrial and the marine realms. An intriguing aspect of the Nanushuk Formation is that it represents a complex of interrelated terrestrial environments and marine environments and sediments such as marine shelf, deltaic, strandplain, fluvial, and alluvial overbank deposits (e.g., Mull et al., 2003; Shimer et al., 2014). With further study, such a mosaic of depositional environments will

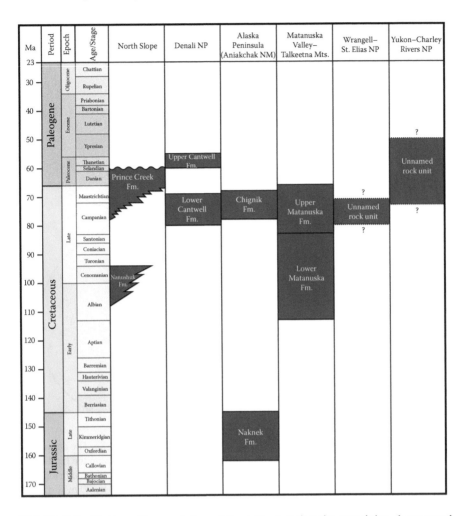

**FIGURE 3.7** Stratigraphic correlation of the known rock units containing documented evidence of dinosaurs in Alaska. That evidence consists of fossil bones, trace fossils, or both. (Compiled from stratigraphic summaries in Detterman, R.L. et al. 1996. *United States Geological Survey Bulletin* 1969-A:1–74; Trop, J.M. and K.D. Ridgway. 2007. Mesozoic and Cenozoic tectonic growth of southern Alaska: A sedimentary basin perspective. In *Tectonic Growth of a Collisional Continental Margin: Crustal Evolution of Southern Alaska*, eds. K.D. Ridgway, J.M. Trop, J.M.G. Glen, and J.M. O'Neill, 55–94. Boulder, CO: Geological Society of America Special Paper 431; Flaig, P.P. et al. 2013. Anatomy, evolution and paleoenvironmental interpretation of an ancient Arctic coastal plain: Integrated paleopedology and palynology from the Upper Cretaceous (Maastrichtian) Prince Creek Formation, North Slope, Alaska, USA. In *New Frontiers in Paleopedology and Terrestrial Paleoclimatology: Paleosols and Soil Surface Analogue Systems*, eds. S.G. Driese and L.C. Nordt, 179–230. SEPM Special Publication 104; Fiorillo, A.R. et al. 2010c. In *New Perspectives on Horned Dinosaurs*, eds. M.J. Ryan, B.J. Chinnery-Allgeier, and D.A. Eberth, 456–477. Bloomington, IN: Indiana University Press.)

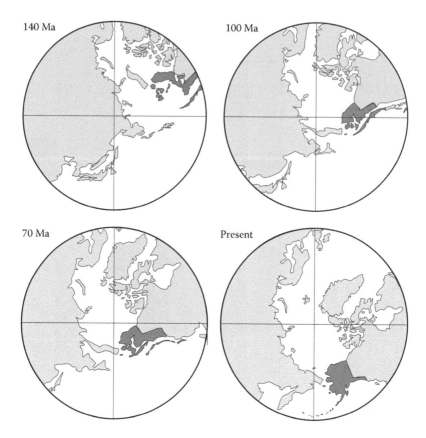

**FIGURE 3.8**   Four time slices representing the tectonic evolution of Alaska. The darker gray represents the major tectonic blocks contributing to the formation of Alaska; the lighter gray represents the surrounding blocks. Notice that by 100 million years ago the land bridge connection between Asia and North America is effectively in place and that by 70 million years ago the northern and southern boundaries of Alaska are near their current latitude. (Redrawn from base maps provided by Lawrence Lawver, Institute of Geophysics, University of Texas.)

almost certainly enlighten us on the details of the relationships between terrestrial biota, their ecosystems, and the paleoclimate of this geologic time interval.

The geology of southern Alaska is dominated by two major collisional events, the arrival of the Wrangellia composite terrane in the Cretaceous and the arrival of the Yakutat terrane in the Cenozoic (Coney et al., 1980; Plafker and Berg, 1994a,b). Within the Wrangellia composite terrane is a thick rock unit called the Matanuska Formation, a largely marine rock unit that captures some of the history of collision of Wrangellia with the rest of Alaska (Trop, 2008). The sediments of the Matanuska Formation are over three kilometers thick and are thought to have been derived from highlands to the north (Jones, 1963; Jones and Grantz, 1967) and deposited in a basin formed as part of the collision between the North American continent to the north and the accreting Wrangellia composite terrane moving in from the south (Trop, 2008). Within the exposures of this rock unit in the Talkeetna Mountains of south-central Alaska, a partial skeleton of a hadrosaur has been reported (Pasch and May, 1997, 2001). The Matanuska

Formation ranges in age from the Early Cretaceous to the Late Cretaceous (Merritt, 1985) and this dinosaur is considered to have come from the Turonian (approximately 94–90 million years ago) portion of the formation (Pasch and May, 1997).

Later Cretaceous dinosaur remains are now documented from around much of Alaska, most notably in the far northern region known as the North Slope along the Colville River (Brouwers et al., 1987; Parrish et al., 1987; Fiorillo and Gangloff, 2000, 2001; Gangloff et al., 2005; Fiorillo, 2008; Fiorillo et al., 2009, 2010b,c; Gangloff and Fiorillo, 2010; Fiorillo and Tykoski, 2012, 2014), in Aniakchak National Monument on the Alaska Peninsula in southwestern Alaska (Fiorillo and Parrish, 2004; Fiorillo et al., 2004), Denali National Park in the Alaska Range of south-central Alaska (Fiorillo et al., 2007, 2011, 2014a,c; Fiorillo and Adams, 2012; Tomsich et al., 2014), Wrangell–St. Elias National Park in the southeastern part of the state (Fiorillo et al., 2012), and along the Yukon River in Yukon–Charley Rivers National Preserve in the eastern interior part of the state (Fiorillo et al., 2014b). The geologic age of the dinosaur-bearing sequences of these rock units within these areas is Campanian–Maastrichtian, or sometime within the window of 84–66 million years ago, though where radiometric dates are available the age is restricted to approximately 72–68 million years ago. Thus, these areas provide a unique opportunity to examine an ancient high-latitude environment at a regional level.

## THE NORTH SLOPE OF ALASKA

The North Slope of Alaska is that part of the state defined as the land that slopes north from the Brooks Range to the Arctic Ocean. It comprises three physiographic provinces: mountains, Arctic Foothills, and Arctic Coastal Plain. At approximately 81,000 square miles, the North Slope is approximately the size of Minnesota, and while the North Slope encompasses about 14% of the geography of Alaska, it is home to only about 1% of the human population.

Underlain by permafrost, much of the coastal landscape when viewed from the air is marked by polygons formed by ice wedges and lakes. The largest river that drains the North Slope is the Colville River, a 350-mile-long waterway that runs from the western Brooks Range down to Harrison Bay, which empties into the Arctic Ocean. Of economic significance are the hydrocarbon reserves that have been discovered in the sedimentary rocks of the North Slope coastal plain.

The uplift of the Brooks Range occurred as a result of the counterclockwise rotation of the Arctic Alaska plate colliding with the rest of Alaska (Figures 3.7 and 3.9). The collision caused a downward bending of the crust to form the Colville Basin. This basin filled with sediments throughout the Late Cretaceous and Tertiary, preserving a package of sedimentary rocks that is several thousand meters thick in some locations (Molenaar et al., 1987).

These sedimentary rocks are within a geologic structure referred to as a foreland basin, and this basin is known as the Colville Basin. Uplift and erosion of the rising Brooks Range during the Cretaceous provided much of the sediments from the west to the basin (Molenaar 1985; Mull, 1985; Moore et al., 1994; Mull et al., 2003; Decker, 2007), while additional sediments entered the basin from the south (Moore et al., 1994). Within the Colville Basin, the Prince Creek Formation is the primary source of polar dinosaur bones in the world. This formation (Figure 3.9)

**FIGURE 3.9** Example of the outcrop pattern for the Prince Creek Formation along the Colville River in northern Alaska. The Prince Creek Formation is the grayish unit comprising much of the outcrop in this photograph. The rock unit is overlain by the Pleistocene Gubik Formation which is the brownish unit in the photograph.

was deposited by a dominantly fluvial system on the Cretaceous Arctic coastal plain (Roehler, 1987; Mull et al., 2003; Flores et al., 2007a,b; Fiorillo et al., 2009, 2010b,c; Flaig et al., 2011, 2013). The Prince Creek Formation was originally subdivided into an older Tuluvak Tongue and a younger Kogosukruk Tongue (Gryc et al., 1951). However, in their extensive review of the Cretaceous stratigraphy of the North Slope, Mull et al. (2003) revised the Prince Creek Formation to include only the former Kogosukruk Tongue and some younger Paleocene strata. By far the best exposures of the Prince Creek Formation are the towering bluffs along the Colville River. The total thickness of the Prince Creek Formation along these bluffs is approximately 450 m (Detterman et al., 1963; Brosge et al., 1966).

In a very general sense, the Prince Creek Formation fines upward from a conglomeratic base to muddier, finer-grained facies at the top (Flores et al., 2007b; Flaig et al., 2011, 2013). In the areas of the river bluffs where the dinosaur bones have been recovered, the Prince Creek Formation is capped by Pliocene/Pleistocene sands of the shallow marine Gubik Formation (Black, 1964) or by a Holocene unconformity.

Biostratigraphic analyses, primarily from fossil pollen and megafloral remains (Frederiksen et al., 1986, 1988, 2002; Parrish and Spicer, 1988; Frederiksen, 1991; Brouwers and de Deckker, 1993; Frederiksen and McIntyre, 2000; Flores et al., 2007b; Brandlen, 2008; Fiorillo et al., 2010c), and isotopic analyses (Conrad et al., 1990; Flaig et al., 2013), indicate the age of the entire Prince Creek Formation ranges from the middle Campanian to middle Paleocene, approximately 79–60 million years ago (Figure 3.7). However, all of the deposits yielding dinosaurs are Early Maastrichtian in age, or approximately 69–70 million years old (Flores et al., 2007b; Fiorillo et al., 2010c; Flaig et al., 2013).

The Prince Creek Formation has several intriguing aspects beyond the role as the primary source of polar dinosaur bones. One of those aspects is the complexity of the depositional system preserved in the sediments, a system that includes a distributary-channel splay complex, interdistributary bays, and the floodplains of delta plains as well as the lower coastal plains. These depositional systems occurred within a high Arctic setting for which there is no modern analog (Flaig et al., 2011, 2013).

Many river channels were in places meandering and anastomosing in others but as a general rule, the trunk channels were meandering while the distributary channels were anastomosing (Fiorillo et al., 2010a; Flaig et al., 2011). The floodplains were generally wet but water levels had seasonal fluctuations between shallow and standing water to dry (Fiorillo et al., 2010b; Flaig et al., 2011, 2013). Soils formed under cool temperate climatic conditions (Flaig et al., 2013). Given the high-latitude depositional setting of these sediments, occasional freezes were possible, but no evidence of cryoturbation has been found within the fossil soils ruling out prolonged periods of freezing (Fiorillo et al., 2010b; Flaig et al., 2013).

North Slope dinosaurs seemed to thrive on a seasonally wet and cool temperate coastal plain, high above the Arctic Circle of the Late Cretaceous. The open woodland areas were coniferous with an angiosperm understory (Spicer, 2003; Spicer and Herman, 2010). Individual dinosaur bones have been recovered from localities deposited within ancient trunk channels, meandering distributaries, and anastomosing distributaries within the Prince Creek Formation. The largest deposits of dinosaur bones, or bonebeds, are found within crevasse splays and low-energy floodplain environments (e.g., Fiorillo et al., 2010b; Gangloff and Fiorillo, 2010). Apparently, seasonal flooding due to snowmelt in combination with alpine permafrost in the Cretaceous Brooks Range created catastrophic flood events on the coastal plain. These seasonal flood events were an effective killing mechanism for the populations of dinosaurs roaming the landscape of this time (Fiorillo et al., 2010b).

The topography of the North Slope during deposition of the Prince Creek Formation was a geographically limited coastal floodplain with the ancestral Brooks Range in one direction and the Arctic Ocean in the other (Flaig et al., 2011). In many ways, the Cretaceous geographic setting for the Prince Creek Formation is similar to the modern Colville River drainage with one exception. Following the Colville River all the way to its delta, cobble-covered river bars are common. Interspersed among these cobbles are colorful cherts derived from later Paleozoic rocks exposed within the Brooks Range, but comparable amounts of very coarse material are absent in the Prince Creek Formation, which presumably had its source near that of the modern Colville River. Was similar coarse material deposited within the Prince Creek Formation only to be removed by later erosion, or is there a depositional basin closer to the Brooks Range that captured the coarsest material eroding out of the rising mountain range—a basin that has yet to be found—and if so, does it contain dinosaurs?

## THE CHIGNIK FORMATION OF ANIAKCHAK NATIONAL MONUMENT

Aniakchak National Monument and Preserve is located on the Alaska Peninsula in an area that is remote even by Alaska standards. This region of southwest Alaska was nicknamed by the geologist and Jesuit priest Bernard Hubbard as the "Cradle of the Storms" (Hubbard, 1931, 1935). Hubbard was one of the first geologists to explore the Alaska Peninsula and was also the first to lead a scientific expedition into what later became Aniakchak National Monument. The Monument comprises almost appproximately 2,400 square kilometers, with an average number of registered visitors reaching only slightly more than 100 people per year. Aniakchak National Monument qualifies as one of the least-visited parks within the United States National Park Service units.

The Park was established in 1978 to recognize the 10-kilometer-wide Aniakchak Caldera, a deep circular feature with walls thousands of feet high formed from the collapse of a magma chamber accompanying a volcanic eruption 3500 years ago (Beget et al., 1992; Pearce et al., 2004; Ringsmuth, 2007). Remarkably, this large feature, which is easily seen in satellite images, was only recognized in 1922, and even then its discovery was not by field sighting but by careful plotting of field survey data in an office (Smith, 1925).

In addition to this remarkable volcanic feature, sedimentary rocks ranging in age from the Late Jurassic (Naknek Formation) to Eocene (Tolstoi Formation) are found throughout the park (Detterman et al., 1981; Wilson et al., 1999). While a variety of invertebrate and plant fossils have been found in many of these units, it is within the Upper Cretaceous Chignik Formation that dinosaurs have been reported (Fiorillo and Parrish, 2004; Fiorillo et al., 2004), and those records are footprints.

The Chignik Formation is part of the Peninsular Terrane, the tectonic block that together with the Wrangellia Terrane formed a superterrane that collided with the more northern blocks of modern Alaska in the Mid- to Late Cretaceous, a wide interval of geologic time approximately 100–66 million years ago. It forms much of southwestern Alaska. The paleogeographic position of this terrane at the time the Chignik Formation was deposited was at approximately its current modern latitude (Hillhouse and Coe, 1994; Lawver et al., 2002). Within the sandstones of this rock unit deposited as collision was occurring are particles of volcanically derived materials. Radiometric date on the mineral zircon derived from igneous rocks indicates that volcanism associated with the collision of the Peninsular Terrane stopped a little more than 90 million years ago. But it is unclear whether these volcanic materials originated near the point of deposition within the Chignik Formation or if these materials were blown in from elsewhere along this collisional boundary.

The Chignik Formation was named by Atwood (1911) for rocks exposed in the vicinity of Chignik Bay, southwest of Aniakchak National Monument. The formation is several hundred meters thick in the area of Chignik Bay, but thickness varies outside the type area, thinning rapidly to the northeast and southwest (Detterman et al., 1996). The rock unit is a cyclic sequence of sediments representing predominantly shallow marine to nearshore marine environments in the lower part and predominantly continental environments in the upper part of the section (Fairchild, 1977; Detterman, 1978; Wahrhaftig et al., 1994; Detterman et al., 1996; Fiorillo and Parrish, 2004; Fiorillo et al., 2004; Figures 3.7 and 3.10). A multitude of dinosaur footprints can be found in the upper Chignik Formation in paleoenvironments representing coastal or terrestrial depositional settings (Fiorillo and Parrish, 2004).

Based on the presence of particular marine bivalves and ammonites, the age of the Chignik Formation is considered to be late Campanian to early Maastrichtian,

**FIGURE 3.10** Sedimentary rocks of the Chignik Formation in Aniakchak National Monument and Preserve in southwestern Alaska. Yoshitsugu Kobayashi (Hokkaido University Museum) is shown in foreground and Paul McCarthy (University of Alaska Fairbanks) in background.

perhaps 72–68 million years old (Detterman et al., 1996), an age that is at least approximately correlative with other dinosaur-bearing units such as the Prince Creek Formation of the North Slope and the Lower Cantwell Formation of Denali National Park (Figure 3.7).

## THE LOWER CANTWELL FORMATION: DENALI NATIONAL PARK

Situated within the 650-kilometer arc of the Alaska Range, a mountain range that extends from the White River in Yukon Territory in the southeast to the area of Lake Clark along the Alaska Peninsula in the southwest is the "crown jewel of the United States National Park System," Denali National Park. Though originally established to help protect wildlife, the park inspires superlatives in response to its abundance of snow-covered peaks, among which is the tallest mountain in North America, Denali. The most dramatic period of mountain building began 5 or 6 million years ago and continues today. Denali is adding to its elevation.

While most visitor attention is focused on the big mountain, or on the abundant modern wildlife, on the north side of the Alaska Range is a rock unit that records an abundance of wildlife from approximately 70 million years ago. That rock unit, the Cantwell Formation, crops out in many areas of the central Alaska Range, including Denali National Park. Divided into two subunits, "Upper Cantwell Formation" and "Lower Cantwell Formation," it is the latter that preserves a rich fossil vertebrate record of latest Cretaceous age, a fossil record that consists of thousands of fossil fish, pterosaur, and dinosaur traces, the most common of these traces being footprints attributable to hadrosaurs (Chapter 4). The Upper Cantwell Formation is dominantly volcanic in origin and is Paleogene in age.

The Lower Cantwell Formation comprises several thousand meters of terrestrial sedimentary deposits (Hickman et al., 1990) that rest upon Jurassic to mid-Cretaceous marine sediments of the Kahiltna assemblage, and on Devonian to Triassic strata of exotic terranes (Wolfe and Wahrhaftig, 1970; Csejtey et al., 1992; Ridgway et al., 1997, 2002; Nokleberg and Richter, 2007; Trop and Ridgway, 2007). The entire Cantwell Formation is intruded by volcanics. The Cantwell Formation fills a geological structural feature known as the Cantwell Basin. The basin, bracketed by two strands of the famous large Denali Fault complex, the Hines Creek Fault to the north, and the McKinley Fault to the south, is 135 km long and up to 35 km wide. While these are the faults that bound the Cantwell Basin, additional faults are commonly found throughout the basin. The result of this structural complexity within the Cantwell Basin makes precise large-scale stratigraphic correlations challenging within the regional outcrop belt for the Cantwell Formation and presents stratigraphic challenges even moving from one side of a mountain to the other.

The Lower Cantwell Formation was assigned an age of late Campanian–early Maastrichtian based on fossil pollen (Ridgway et al., 1997). Recent isotopic dates from bentonites, a clay mineral that forms from the weathering of volcanic ash, corroborate these pollen age determinations, as the bentonites yield ages of approximately 71.5–69.5 million years (Tomsich et al., 2014; Salazar-Jaramillo et al., 2016; Figure 3.7). Hillhouse and Coe (1994) place the Cantwell Basin at a paleolatitude of approximately 71° north, indicating that the Lower Cantwell Formation was within the ancient Arctic at the time of deposition.

Large parts of the Lower Cantwell Formation are dominated by fine-grained sandstones and siltstones interbedded with mudstones and relatively thick successions of coaly shale. Large fluvial channels are preserved as multistoried bodies of massive conglomerate, pebbly sandstone and coarse- to medium-grained sandstone in successions up to 5 m thick, representing sandy, braided trunk channels in an axial drainage system within the lower Cantwell landscape (Fiorillo et al., 2014a; Figure 3.11).

Overbank deposits are represented by medium- to fine-grained sandstones, siltstones, mudstones, and coaly shales (Figure 3.11). Interbedded sandstones, siltstones, and mudstones represent splay deposits distal from active distributary channels. Thicker siltstone and mudstone deposits are interpreted as small lakes or ponds, while coaly shales are interpreted as backswamps, which were poorly drained and infrequently received inputs of fine-grained clastic material (Fiorillo et al., 2014a; Tomsich et al., 2014). Deposition varied from unconfined shallow overland flow inundating floodplain vegetation and rapidly decelerating due to obstruction of flow by plants, to a more mature floodplain forest on the distal part of an alluvial fan or on an inactive alluvial fan lobe that occasionally received distal sediment deposition (Fiorillo et al., 2015).

## AN UNNAMED ROCK UNIT: WRANGELL–ST. ELIAS NATIONAL PARK

At over 54,000 square kilometers, an area larger than Switzerland, Wrangell–St. Elias National Park, located in southern Alaska, is the largest national park in the United States. The mountains of Wrangell–St. Elias rank among the tallest in North America, earning the region the nickname the Crown of the Continent. The landscape is vast and varied. The park includes the highest coastal range in the world,

**FIGURE 3.11** Exposures of the Cantwell Formation in Denali National Park and Preserve, Alaska. The gray rocks in the foreground are part of the sedimentary sequence comprising the Lower Cantwell Formation.

with towering Mount St. Elias almost 5500 meters in elevation along the Gulf of Alaska. This remarkable topographic relief is the result of the second major collisional event defining the geology of southern Alaska, the accretion of the Yakutat terrane. The dynamic tectonic activity of the region produced the Wrangell Volcanic Field with active volcanoes including Mounts Blackburn, Wrangell, and Sanford dominating the landscape. In addition, the park includes the Bagley Icefield, North America's largest subpolar icefield.

The coast of what is now Wrangell–St. Elias National Park has seen the likes of Vitus Bering, Grigory Shelikov, James Cook, and George Vancouver, to name just a few of the sea captains who explored the area. While these explorers spent time along the coast, by the latest part of the 1800s reports of rich ore deposits from farther inland were being confirmed, which led to many geologic exploration expeditions throughout the region. With modern studies, it is now recognized that the geologic bedrock of the park comprises several tectonic terranes, all of which, starting around 200 million years ago, moved northward through time to their point of collision.

Discoveries of copper and gold deposits caused prospectors to pour into the region, followed by detailed geologic surveys. The first was led by Lieutenant Henry Allen of the U.S. Cavalry (1887). Allen's geologic observations focused on the obvious stunning features of the landscape such as the active volcanoes, glaciers, and ore deposits, tucked away in the northeastern part of the park. Where the gold deposits were played out by the early 1900s is an unnamed sequence of latest Cretaceous sedimentary rocks. These unnamed rocks are part of the Wrangellia Terrane (e.g., Winkler, 2000), a composite terrane containing various Mesozoic-aged rocks. During deposition of Triassic rocks contained within this terrane, Wrangellia was several thousand kilometers south of its current latitude. However, by the Late Cretaceous

**FIGURE 3.12** Exposures of unnamed Cretaceous rocks in Wrangell–St. Elias National Park and Preserve in southeastern Alaska. The unnamed unit is exposed in the rolling landscape in the foreground.

Wrangellia was near its current latitude (Hillhouse and Coe, 1994; Winkler, 2000; Stamatakos et al., 2001), indicating that the dinosaurs discovered there lived in the higher latitudes (Fiorillo et al., 2012).

These unnamed fluvial rocks, representing Cretaceous rivers and floodplains, were laid down in the Nutzotin structural basin, which formed at the northern edge of the Wrangellia terrane (Manuszak et al., 2007). The geographic extent of these rocks is relatively small (Figure 3.12), and exposures are very limited, bounded in all directions by faults. The poor exposures and structural complexities of the area are presumably what has left this rock sequence without a proper name. Underlying these nonmarine rocks is the largely volcanic Chisana Formation, a rock unit comprised of fragmental volcanics that were deposited in either a submarine or subaerial setting.

In their study of the long-term Mesozoic (Late Triassic–Late Cretaceous) sedimentary basin development of the Wrangellia composite terrane, Trop et al. (2002) examined in detail the evolution of the Wrangell Mountains basin and the Nutzotin Basin. Using $^{40}$Ar/$^{39}$Ar ages and palynological data, they correlated intrabasinal and deformational events between the two basins. This resulted in their tentative placement of this unnamed sequence of clastic sedimentary rocks that yielded the dinosaur footprints now recorded from the unit (Fiorillo et al., 2012) in the latest Cretaceous, somewhere between 80 and 66 million years ago (Figure 3.7; Trop et al., 2002). The conclusion by Trop et al. (2002) corroborated earlier work by geologists of the United States Geological Survey (Richter, 1976; Richter et al., 2006).

Given the tectonic setting of the Nutzotin Basin, a small basin undergoing active folding, coarse conglomerates containing cobbles and pebbles of pre-existing rocks from beyond the depositional basin dominate the available exposures. These conglomerate beds are often over 20 meters in thickness and represent large-scale river channel systems. Given this high-energy depositional environment, which presumably broke apart most fossil material prior to deposition, the fossils recorded from these conglomerates are restricted to small pieces of gymnosperm wood (Fiorillo et al., 2012).

Within the finer-grained overbank layers of this unnamed unit, which represent much lower-energy depositional environments, plant debris is much more common, including upright tree trunks and three-dimensionally preserved ferns (Fiorillo et al., 2012). These softer overbank deposits are also much less abundant than the more weather-resistant conglomeratic beds. An abundance of charcoal in these overbank deposits suggests that fire was a common occurrence within the ecosystem preserved in these rocks. The record of regularly occurring fires combined with the abundance of high-energy deposition suggests a tectonically dynamic landscape prone to ecological disturbance, a somewhat different environmental setting than that recorded in the correlative Prince Creek Formation or the Lower Cantwell Formation. Also within these fine-grained rocks, dinosaur tracks were found, which further established the widespread nature and the ecosystem capable of supporting these animals (Fiorillo et al., 2012). While the tectonic history of this complex and geologically dynamic region is important in order to understand the large-scale processes that formed our Earth, further work may also shed light on the adaptability of life during the Cretaceous on the dynamic terrestrial ecosystem along this convergent boundary between Alaska and Wrangellia.

## MORE UNNAMED ROCKS IN THE ALASKAN INTERIOR: YUKON–CHARLEY RIVERS NATIONAL PRESERVE

For fans of the popular writings of Jack London and Robert Service, the landscape along Yukon River drainage of Alaska and the Yukon Territory of Canada inspired such literary works as *The Call of the Wild*, *White Fang*, and *The Spell of the Yukon*. On the Alaskan side of the boundary between the United States and Canada is a national park unit known as Yukon–Charley Rivers National Preserve. The park encompasses some 10,000 square kilometers along some 185 kilometers of the Yukon River.

Glaciation, which featured so prominently in the recent geologic past of Alaska, bypassed most of the area around Yukon–Charley Rivers National Preserve. The park has a rich deeper geologic history with a prominent structural feature known as the Tintina Fault, a strike-slip fault that extends from British Columbia into central Alaska. The fault is interpreted as an extension of the fault that runs along the Rocky Mountains in the continental United States, making this one of the longest linear geologic features in North America. Lateral movement along the fault zone in Alaska is thought to have been on the order of several hundred kilometers (Gabrielse, 1985; Gabrielse et al., 2006; Saltus, 2010), and major movements likely ended in the Late Cretaceous but minor movements may have occurred as recently as the Paleogene (Gabrielse, 1985; Dover, 1994; Gabrielse et al., 2006; Till et al., 2007; Saltus, 2010).

Yukon–Charley Rivers National Preserve is known to contain an abundant and diverse fossil record within the sedimentary rocks in the park. These fossil-bearing rock strata range in age from the Proterozoic to the Pleistocene, effectively

**FIGURE 3.13** Example of outcrop pattern of unnamed Cretaceous–Tertiary rocks in Yukon–Charley Rivers National Preserve located in east-central Alaska.

covering the last billion years of life on Earth (Allison, 1988; Dover, 1994). Within this thick sequence of sedimentary rocks is another unnamed fluvial sedimentary unit (Figure 3.13). Though poorly exposed at the surface, the unnamed rock unit is geographically expansive, and it is far from trivial in volume, being up to a couple of thousand meters thick. The unnamed rock unit is mapped simply as unit TKs or TKd (Brabb and Churkin, 1969; Foster, 1976; Dover, 1994). The former label has most often been used by subsequent workers (Dover and Miyaoka, 1988; Miyaoka, 1990; Foster, 1992; Van Kooten et al., 1997; Johnsson, 2000). TKs is made up of dominantly fluviatile rocks that are generally poorly indurated and consist of con-glomerates, sandstones, siltstones, and coal, and contain significant Late Cretaceous and Paleogene flora (Brabb and Churkin, 1969). The TKs unit occurs within the fault zone and overlies older sedimentary rocks to the north of the Tintina fault (Howell et al., 1992; Van Kooten et al., 1997).

Some workers have suggested that the TKs sedimentary unit was deposited in an intracontinental basin that resulted from extension of the crust starting in the latest Cretaceous (Van Kooten et al., 1997; Johnsson, 2000). The structural and stratigraphic relationships of the region suggest that this unit is Late Cretaceous or younger in age, but due to the sporadic nature of the outcrop pattern and the difficulty in access to individual outcrops, the geologic age of the sediments in this basin as well as the stratigraphic relationships between rock exposures is not well understood.

Early reports of megafloral remains suggested an Eocene age (Prindle, 1913), while Martin (1926) assigned the rocks to the Late Cretaceous. Still, Hollick (1930, 1936) sampled rocks in the Yukon–Charley rivers region and favored an age younger than the Cretaceous, and assigned the rocks to a Paleogene age. Because he was unable to resolve the issue, during Mertie's later work (1942) he accepted the nature of the mixed age data available and he published his floral lists as matter of record for future generations.

Subsequent workers have also been unable to resolve this stratigraphic issue and have chosen to carry through the concept that this unnamed rock unit spans the latest Cretaceous and Paleogene (Figure 3.7; Brabb and Churkin, 1969; Foster, 1976; Dover and Miyaoka, 1988; Foster and Igarashi, 1989; Johnsson, 2000). However, a recent study of the TKs unit yielded the first evidence of dinosaur remains from the region and concluded that the lower part of the unnamed rock unit is rooted in the latest Cretaceous (Fiorillo et al., 2014b). That the data for the latest Cretaceous age are derived from only within the Tintina Fault zone suggests that the dynamics of movement along the fault brought the basal-most deposits of the TKs unit to the surface to be exposed (Fiorillo et al., 2014b), which then raises questions such as how extensive is this type of movement within the Tintina Fault zone, and what other discoveries can be made within this fault zone that might add further insights into the Cretaceous terrestrial ecosystem in this region?

The available chronostratigraphic data suggest that the part of the TKs section that contains dinosaur tracks is correlative with the Prince Creek Formation. Of interest in that possibility is not only the occurrence of dinosaur tracks, but the paleobotany within these rocks further suggests that from a Late Cretaceous regional perspective, the ecosystem was heterogeneous (Fiorillo et al., 2014b).

## SUMMARY

The geologic history of Alaska is rich and complex. Except for a relatively small wedge of bedrock in the northeastern part of the state, the region now known as Alaska is an amalgamation of tectonic blocks building the landscape over time. Caught between the tectonic pressures caused during the opening of the Arctic Basin pushing the northern Alaska block south and the tectonic pressures of the Pacific Plate pushing southern Alaskan blocks north, by the middle of the Cretaceous Period Alaska was effectively at its current latitude. As a result of these tectonic movements, a vast heterogeneous terrestrial ecosystem was in place in the ancient Arctic during the Cretaceous, an ecosystem capable of supporting large numbers of dinosaurs. Further, these Cretaceous dinosaurs of Alaska were residents of the ancient high latitudes rather than having lived somewhere in the southern latitudes only to have been highjacked and moved north by later tectonic movements.

## REFERENCES

Adams, T.L. 2009. Deposition and taphonomy of the Hound Island Late Triassic vertebrate fauna: Fossil preservation within subaqueous gravity flows. *Palaios* 24:603–615.

Allen, H.T. 1887. *Report of an Expedition to the Copper, Tanana, and Koyukuk Rivers, in the Territory of Alaska, in the Year 1885*. Washington, DC: Government Printing Office, 172pp.

Allison, C.W. 1988. Paleontology of Late Proterozoic and Early Cambrian rocks of east-central Alaska. *United States Geological Survey Professional Paper* 1449:1–50.

Amato, J.M. and T.L. Pavlis. 2010. Detrital zircon ages from the Chugach terrane, southern Alaska, reveal multiple episodes of accretion and erosion in a subduction complex. *Geology* 38:459–462.

Atwood, W.W. 1911. Geology and mineral resources of parts of the Alaska Peninsula. *United States Geological Survey Bulletin* 467:1–137.

Beget, J.E., O. Mason, and P. Anderson. 1992. Age, extent and climatic significance of the c. 3400 BP Aniakchak tephra, western Alaska, USA. *Holocene* 2:51–56.

Black, R.F. 1964. Gubik Formation of Quaternary age in northern Alaska. *United States Geological Survey Professional Paper* 302-C:59–91.

Blieck, A.R.M., V.N. Karatajute-Talimaa, and E. Mark-Kurik. 2002. Upper Silurian and Devonian heterostracan pteraspidomorphs (Vertebrata) from Severnaya Zemlya (Russia): A preliminary report with biogeographical and biostratigraphical implications. *Geodiversitas* 24:805–820.

Blodgett, R.B. and G.D. Stanley, Jr., eds. 2008. *The Terrane Puzzle: New Perspectives on Paleontology and Stratigraphy from the North American Cordillera.* Boulder, CO: Geological Society of America Special Paper 442.

Brabb, E.E. and M. Churkin, Jr. 1969. Geologic map of the Charley River Quadrangle, east-central Alaska. *United States Geological Survey Miscellaneous Geologic Investigations Map-573, 1:250,000.*

Brandlen, E. 2008. *Paleoenvironmental reconstruction of the Late Cretaceous (Maastrichtian) Prince Creek Formation, near the Kikak–Tegoseak dinosaur quarry, North Slope, Alaska. MS thesis,* University of Alaska–Fairbanks.

Broad, D.S. and A.C. Lenz. 1972. A new Upper Silurian species of *Vernonaspis* (Heterostraci) from Yukon Territory, Canada. *Journal of Paleontology* 46:415–420.

Brosge, W.P., C.L. Whimington, and R.H. Morris. 1966. Geology of the Umiat–Maybe Creek Region, Alaska. *United States Geological Survey Professional Paper* 303-H:548–570.

Brouwers, E.M. and P. De Deckker. 1993. Late Maastrichtian and Danian ostracode faunas from northern Alaska: Reconstructions of environment and paleogeography. *Palaios* 8:140–154.

Brouwers, E.M., W.A. Clemens, R.A. Spicer, T.A. Ager, L.D. Carter, and W.V. Sliter. 1987. Dinosaurs on the North Slope, Alaska: High latitude, latest Cretaceous environments. *Science* 237:1608–1610.

Burchfiel, B.C., D.S. Cowan, and G.A. Davis. 1992. Tectonic overview of the Cordilleran orogen in the western U.S. In *The Cordilleran Orogen: Conterminous U.S., Geology of North America G-3,* eds. B.C. Burchfiel, P.W. Lipman, and M.L. Zoback, 407–480. Boulder, CO: Geological Society of America.

Callaway, J.M. and J.A. Massare. 1989. Geographic and stratigraphic distribution of the Triassic Ichthyosauria (Reptilia, Diapsida). *Neues Jahrbuch für Geologie und Palaeontologie. Abhandlungen* 178:37–58.

Campbell, L.J. 1994. The Terrible Lizards. *Alaska Geographic* 21:24–37.

Clark, S.H.B. 1973. The McHugh Complex of south-central Alaska. *United States Geological Survey Bulletin* 1372-D:1–11.

Coney, P.J. and C.A. Evenchick. 1994. Consolidation of the American Cordilleras. *Journal of South American Earth Sciences* 7:241–262.

Coney, P.J. D.L. Jones, and J.W.H. Monger. 1980. Cordilleran suspect terranes. *Nature* 288:329–333.

Conrad, J.E., E.H. McKee, and B.D. Turrin. 1990. Age of Tephra Beds at the Ocean Point Dinosaur Locality, North Slope, Alaska, based on K-Ar and $^{40}$Ar/$^{39}$Ar analyses. *United States Geological Survey Bulletin* 1990-C:1–12.

Conyers, L. 1978. Letters, notes, and comments. *Alaska Magazine* 44:30.

Csejtey, B., M.W. Mullen, D.P. Cox, and G.D. Stricker. 1992. Geology and geochronology of the Healy quadrangle, south-central Alaska. *United States Geological Survey Miscellaneous Investigation Series* I-1961, 63pp, 2 plates, scale 1:250,000.

Daeschler, E.B., N.H. Shubin, and F.A. Jenkins, Jr. 2006. A Devonian tetrapod-like fish and the evolution of the tetrapod body plan. *Nature* 440:757–763.

Decker, P.L. 2007. Brookian Sequence Stratigraphic Correlations, Umiat Field to Milne Point Field, West-Central North Slope, Alaska. *Alaska Department of Natural Resources Preliminary Interpretive Report* 2007-2, 21pp, 1 map.

Denison, R.H. 1963. New Silurian Heterostraci from Southeastern Yukon. *Fieldiana, Geology* 14:105–141.

Detterman, R.L. 1978. Interpretation of depositional environments in the Chignik Formation, Alaska Peninsula. *United States Geological Survey, Circular* 772-B:62–63.

Detterman, R.L., R.S. Bickel, and G. Gryc. 1963. Geology of the Chandler River Region, Alaska. *United States Geological Survey Special Paper* 303-E:223–324.

Detterman, R.L., J.E. Case, J.W. Miller, F.H. Wilson, and M.E. Yount. 1996. Stratigraphic framework of the Alaska Peninsula. *United States Geological Survey Bulletin* 1969-A:1–74.

Detterman, R.L., T.P. Miller, M.E. Yount, and F.H. Wilson. 1981. Geologic map of the Chignik and Sutwik Island Quadrangles, Alaska. *United States Geological Survey Miscellaneous Investigations Series*, Map I-1229, 1:250,000.

Dewey, J.F. and J.M. Bird. 1970. Mountain belts and the new global tectonics. *Journal of Geophysical Research* 75:2625–2647.

Dineley, D.L. and E.J. Loeffler. 1976. Ostracoderm faunas of the Delorme and associated Siluro-Devonian Formations North West Territories Canada. *Special Papers in Palaeontology* 18:1–214.

Dover, J.H. 1994. Geology of part of east-central Alaska. In *The Geology of Alaska: The Geology of North America G-1, Decade of North American Geology*, eds. G. Plafker and H.C. Berg, 153–204. Boulder, CO: Geological Society of America.

Dover, J.H. and R.T. Miyaoka. 1988. Reinterpreted geologic map and fossil data, Charley River Quadrangle, East-Central Alaska. *United States Geological Survey Miscellaneous Field Studies Map* Map MF-2004, 1:250,000.

Druckenmiller, P.S. N. Kelley, M.T. Whalen, C. McRoberts, and J.G. Carter. 2014. An Upper Triassic (Norian) ichthyosaur (Reptilia, Ichthyopterygia) from northern Alaska and dietary insight based on gut contents. *Journal of Vertebrate Paleontology* 34:1460–1465.

Druckenmiller, P.S. and E.E. Maxwell. 2014. A Middle Jurassic (Bajocian) ophthalmosaurid (Reptilia Ichthyosauria) from the Tuxedni Formation, Alaska and the early diversification of the clade. *Geological Magazine* 151:41–48.

Dunkle, D.H. 1964. United States Geological Survey Examine and Reports A-63-22, locality 63ACn1514. Unpublished.

Elliott, D.K. 1983. New Pteraspididae (Agnatha, Heterostraci) from the Lower Devonian of Northwest Territories, Canada. *Journal of Vertebrate Paleontology* 2:389–406.

Elliott, D.K. 1984. A new subfamily of the Pteraspididae (Agnatha, Heterostraci) from the Upper Silurian and Lower Devonian of Arctic Canada. *Journal of Paleontology* 27:169–197.

Elliott, D.K. and D.L. Dineley. 1983. New species of *Protopteraspis* (Agnatha, Heterostraci) from the Lower Devonian of Northwest Territories, Canada. *Journal of Paleontology* 56:474–494.

Fairchild, D.T. 1977. Paleoenvironments of the Chignik Formation, Alaska Peninsula. Master's thesis, University of Alaska, Fairbanks, 168 pp. (Unpublished).

Fiorillo, A.R. 2006. Review of the dinosaur record of Alaska with comments regarding Korean dinosaurs as comparable high-latitude fossil faunas. *Journal of the Paleontological Society of Korea* 22:15–27.

Fiorillo, A.R. 2008. On the occurrence of exceptionally large teeth of *Troodon* (Dinosauria: Saurischia) from the Late Cretaceous of northern Alaska. *Palaios* 23:322–328.

Fiorillo, A.R. and T.L. Adams. 2012. A therizinosaur track from the Lower Cantwell Formation (Upper Cretaceous) of Denali National Park, Alaska. *Palaios* 27:395–400.

Fiorillo, A.R., T.L. Adams, and Y. Kobayashi. 2012. New sedimentological, palaeobotanical, and dinosaur ichnological data on the palaeoecology of an unnamed Late Cretaceous rock unit in Wrangell–St. Elias National Park and Preserve, Alaska, USA. *Cretaceous Research* 37:291–299.

Fiorillo, A.R., P. Armato, and R. Kucinski. 2004. Wandering rocks in Kenai Fjords National Park. *Alaska Park Science* 3:21–23.

Fiorillo, A.R., M. Contessi, Y. Kobayashi, and P.J. McCarthy. 2014a. Theropod tracks from the Lower Cantwell Formation (Upper Cretaceous) of Denali National Park, Alaska, USA with comments on theropod diversity in an ancient, high-latitude terrestrial ecosystem. In *Tracking Dinosaurs and Other Tetrapods in North America*, eds. M. Lockley and S.G. Lucas, 429–439, Albuquerque, NM: *New Mexico Museum of Natural History and Science Bulletin* 62.

Fiorillo, A.R., P.L. Decker, D.L. LePain, M. Wartes, and P.J. McCarthy. 2010a. A probable Neoceratopsian Manus Track from the Nanushuk Formation (Albian, Northern Alaska). *Journal of Iberian Geology* 36:165–174.

Fiorillo, A.R., F. Fanti, C. Hults, and S.T. Hasiotis. 2014b. New ichnological, paleobotanical and detrital zircon data from an unnamed rock unit in Yukon–Charley Rivers National Preserve (Cretaceous: Alaska): Stratigraphic implications for the region. *Palaios* 29:16–26.

Fiorillo, A.R. and R.A. Gangloff. 2000. Theropod teeth from the Prince Creek Formation (Cretaceous) of northern Alaska, with speculations on arctic dinosaur paleoecology. *Journal of Vertebrate Paleontology* 20:675–682.

Fiorillo, A.R. and R.A. Gangloff. 2001. The caribou migration model for Arctic hadrosaurs (Ornithischia: Dinosauria): A reassessment. *Historical Biology* 15:323–334.

Fiorillo, A.R., S.T. Hasiotis, and Y. Kobayashi. 2014c. Herd structure in Late Cretaceous polar dinosaurs: A remarkable new dinosaur tracksite, Denali National Park, Alaska, USA. *Geology* 42:719–722.

Fiorillo, A.R., S.T. Hasiotis, Y. Kobayashi, B.H. Breithaupt, and P.J. McCarthy. 2011. Bird tracks for the Upper Cretaceous Cantwell Formation of Denali National Park, Alaska, USA: A new perspective on ancient polar vertebrate biodiversity. *Journal of Systematic Palaeontology* 9:33–49.

Fiorillo, A.R., Y. Kobayashi, P.J. McCarthy, T.C. Wright, and C.S. Tomsich. 2015. Reports of pterosaur tracks from the Lower Cantwell Formation (Campanian-Maastrichtian) of Denali National Park, Alaska, USA, with comments about landscape heterogeneity and habitat preference. *Historical Biology* 27:672–683.

Fiorillo, A.R., P.J. McCarthy, B. Breithaupt, and P. Brease. 2007. Dinosauria and fossil Aves footprints from the Lower Cantwell Formation (latest Cretaceous), Denali Park and Preserve, Alaska. *Alaska Park Science* 6:41–43.

Fiorillo, A.R., P.J. McCarthy, and P.P. Flaig. 2010b. Taphonomic and sedimentologic Interpretations of the dinosaur-bearing upper Cretaceous strata of the Prince Creek Formation, Northern Alaska: Insights from an ancient high-latitude terrestrial ecosystem. *Palaeogeography, Palaeoclimatology, Palaeoecology* 295:376–388.

Fiorillo, A.R., P.J. McCarthy, P.P. Flaig, E. Brandlen, D.W. Norton, P. Zippi, L. Jacobs, and R.A. Gangloff. 2010c. Paleontology and paleoenvironmental interpretation of the Kikak-Tegoseak Quarry (Prince Creek Formation: Late Cretaceous), northern Alaska: A multi-disciplinary study of a high-latitude ceratopsian dinosaur bonebed. In *New Perspectives on Horned Dinosaurs*, eds. M.J. Ryan, B.J. Chinnery-Allgeier, and D.A. Eberth, 456–477. Bloomington, IN: Indiana University Press.

Fiorillo, A.R. and J.T. Parrish. 2004. The first record of a Cretaceous dinosaur from western Alaska. *Cretaceous Research* 25:453–458.

Fiorillo, A.R. and R.S. Tykoski. 2012. A new species of centrosaurine ceratopsid *Pachyrhinosaurus* from the North Slope (Prince Creek Formation: Maastrichtian) of Alaska. *Acta Palaeontologica Polonica* 57:561–573.

Fiorillo, A.R. and R.S. Tykoski. 2014. A diminutive new tyrannosaur from the top of the world. *PLoS ONE* 9(3):e91287. doi:10.1371/journal.pone.0091287.

Fiorillo, A.R., R.S. Tykoski, P.J. Currie, P.J. McCarthy, and P. Flaig. 2009. Description of two partial Troodon braincases from the Prince Creek Formation (Upper Cretaceous), North Slope Alaska. *Journal of Vertebrate Paleontology* 29:178–187.

Flaig, P., P.J. McCarthy, and A.R. Fiorillo. 2011. A tidally influenced, high-latitude alluvial/coastal plain: The Late Cretaceous (Maastrichtian) Prince Creek Formation, North Slope, Alaska. In *From River to Rock Record: The Preservation of Fluvial Sediments and Their Subsequent Interpretation*, eds. C. North, S. Davidson, and S. Leleu, 233–264. Tulsa, OK: SEPM Special Publication 97.

Flaig, P.P., P.J. McCarthy, and A.R. Fiorillo. 2013. Anatomy, evolution and paleoenvironmental interpretation of an ancient Arctic coastal plain: Integrated paleopedology and palynology from the Upper Cretaceous (Maastrichtian) Prince Creek Formation, North Slope, Alaska, USA. In *New Frontiers in Paleopedology and Terrestrial Paleoclimatology: Paleosols and Soil Surface Analogue Systems*, eds. S.G. Driese and L.C. Nordt, 179–230. Tulsa, OK: SEPM Special Publication 104.

Flores, R.M., M.D. Myers, D.W. Houseknecht, G.D. Stricker, D.W. Brizzolara, T.J. Ryherd, and K.I. Takahashi. 2007a. Stratigraphy and Facies of Cretaceous Schrader Bluff and Prince Creek Formations in Colville River Bluffs, North Slope, Alaska. *United States Geological Survey Professional Paper* 1748:1–52.

Flores, R.M., G.D. Stricker, P.L. Decker, and M.D. Myers. 2007b. Sentinel Hill Core Test 1: Facies Descriptions and Stratigraphic Reinterpretations of the Prince Creek and Schrader Bluff Formations, North Slope, Alaska. *United States Geological Survey Professional Paper* 1747:1–31.

Foster, H.L. 1976. Geologic map of the Eagle quadrangle, Alaska. *United States Geological Survey Miscellaneous Investigations Series*, Map I-922, scale 1:250,000, 1 sheet.

Foster, H.L. 1992. Geologic map of the Eastern Yukon–Tanana region, Alaska. *United States Geological Survey Open-File Report 92-313*, p. 26, scale 1:250,000, 1 sheet.

Foster, H.L. and Y. Igarashi. 1989. Fossil pollen from nonmarine sedimentary rocks of the eastern Yukon–Tanana region, east-central Alaska. *United States Geological Survey Bulletin* 1946:11–20.

Frederiksen, N.O. 1991. Pollen Zonation and Correlation of Maastrichtian Marine Beds and Associated Strata, Ocean Point Dinosaur Locality, North Slope, Alaska. *United States Geological Survey Bulletin* 1990-E:1–24.

Frederiksen, N.O, T.A. Ager, and L.E. Edwards. 1986. Comment on "Early Tertiary marine fossils from northern Alaska: Implications for Arctic Ocean paleogeography and faunal evolution." *Geology* 14:802–803.

Frederiksen, N.O., T.A. Ager, and L.E. Edwards. 1988. Palynology of Maastrichtian and Paleocene rocks, lower Colville River region, North Slope, Alaska. *Canadian Journal of Earth Sciences* 25:512–527.

Frederiksen, N.O. and D.J. McIntyre. 2000. Palynomorph Biostratigraphy of Mid(?) Campanian to Upper Maastrichtian Strata Along the Colville River, North Slope of Alaska. *United States Geological Survey Open-File Report* 00–493:1–36.

Frederiksen, N.O., D.J. McIntyre, and T.P. Sheehan. 2002. Palynological Dating of Some Upper Cretaceous to Eocene Outcrop and Well Samples from the Region Extending from the Easternmost Part of NPRA in Alaska to the Western Part of ANWR, North Slope of Alaska. *United States Geological Survey Open-File Report* 02–405:1–37.

Gabrielse, H. 1985. Major dextral transcurrent displacements along the Northern Rocky Mountain Trench and related lineaments in north-central British Columbia. *Geological Society of America Bulletin* 96:1–14.

Gabrielse, H., D.C. Murphy, and J.K. Mortensen. 2006. Cretaceous and Cenozoic dextral orogen-parallel displacements, magmatism, and paleogeography, north central Canadian Cordillera. In *Paleogeography of the North American Cordillera: Evidence for and against Large-Scale Displacements*, eds. W.J. Haggart, J.R. Enkin, and H.J.W. Monger, 255–276. St. John's, NL: Geological Association of Canada Special Paper 46.

Gangloff, R.A. 1994. The record of Cretaceous dinosaurs in Alaska: An overview. In *1992 Proceedings International Conference on Arctic Margins*, eds. D.K. Thurston and K. Fujita, 399–404. Anchorage, AK: Outer Continental Shelf Study, Mineral Management Service 94-0040.

Gangloff, R.A. 1998. Newly discovered dinosaur trackways from the Cretaceous Chandler Formation, National Petroleum Reserve-Alaska. *Journal of Vertebrate Paleontology* 18(Suppl 3):45A.

Gangloff, R.A. and A.R. Fiorillo. 2010. Taphonomy and paleoecology of a bonebed from the Prince Creek Formation, North Slope, Alaska. *Palaios* 25:299–317.

Gangloff, R.A., A.R. Fiorillo, and D.W. Norton. 2005. The first Pachycephalosaurine (Dinosauria) from the Paleo-Arctic and its paleogeographic implications. *Journal of Paleontology* 79:997–1001.

Gradstein, F., J. Ogg, and A. Smith, eds. 2004. *A Geologic Time Scale 2004*. Cambridge: Cambridge University Press.

Gryc, G., W.W. Patton, and T.G. Payne. 1951. Present stratigraphic nomenclature of northern Alaska. *Journal of the Washington Academy of Sciences* 41:159–167.

Guthrie, R.D. 1990. *Frozen Fauna of the Mammoth Steppe: The Story of Blue Babe*. Chicago, IL: University of Chicago Press.

Guthrie, R.D. 2003. Rapid body size decline in Alaskan Pleistocene horses before extinction. *Nature* 426:169–171.

Guthrie, R.D. and J.V. Matthews. 1971. The Cape Deceit fauna—Early Pleistocene mammalian assemblage from the Alaskan Arctic. *Quaternary Research* 1:474–510.

Hickman, R.G., K.W. Sherwood, and C. Craddock. 1990. Structural evolution of the early Tertiary Cantwell basin, south-central Alaska. *Tectonics* 9:1433–1449.

Hillhouse, J.W. and R.S. Coe. 1994. Paleomagnetic data from Alaska. In *The Geology of Alaska. The Geology of North America G-1*, eds. G. Plafker and H.C. Berg, 797–812. Boulder, CO: Geological Society of America.

Hollick, C.A. 1930. The Upper Cretaceous floras of Alaska. *United States Geological Survey Professional Paper* 159:1–123.

Hollick, C.A. 1936. The Tertiary floras of Alaska. *United States Geological Survey Professional Paper* 182:1–185.

Howell, D.G., M.J. Johnsson, M.B. Underwood, L., Huafu, and J.W. Hillhouse. 1992. Tectonic evolution of the Kandik Region, east-central Alaska: Preliminary interpretations. *United States Geological Survey Bulletin* 1999:127–140.

Hubbard, B.R. 1931. A world inside a mountain. *National Geographic* 60:319–345.

Hubbard, B.R. 1935. *Cradle of the Storms*. New York: Dodd, Mead & Company.

Johnsson, M.J. 2000. Tectonic assembly of East-Central Alaska: Evidence from Cretaceous–Tertiary sandstones of the Kandik River Terrane. *Geological Society of America Bulletin* 112:1023–1042.

Jones, D.L. 1963. Upper Cretaceous (Campanian and Maestrichtian) ammonites from southern Alaska. *United States Geological Survey Professional Paper* 432:1–53.

Jones, D.L. and A. Grantz. 1967. Cretaceous ammonites from the lower part of the Matanuska Formation, southern Alaska. *United States Geological Survey Professional Paper* 547:1–49.

Lawver, L.A., A. Grantz, and L.M. Gahagan. 2002. Plate kinematic evolution of the present Arctic Region since the Ordovician. In *Tectonic Evolution of the Bering Shelf-Chukchi Sea-Arctic Margin and Adjacent Landmasses*, eds. E.L. Miller, A. Grantz, and S.L. Klemperer, 333–358. Boulder, CO: Geological Society of America Special Paper 360.

Lawver, L.A. and C.R. Scotese. 1990. A review of tectonic models for the evolution of the Canada Basin. In *The Arctic Ocean Region. The Geology of North America, Vol. L*, eds. A. Grantz, L. Johnson, and J.F. Sweeney, 593–618. Boulder, CO: Geological Society of America.

Manuszak, J.D., K.D. Ridgway, J.M. Trop, and G.E. Gehrels. 2007. Sedimentary record of the tectonic growth of a collisional continental margin: Upper Jurassic-Lower Cretaceous Nutzotin Mountains sequence, eastern Alaska Range, Alaska. In *Tectonic Growth of a Collisional Continental Margin: Crustal Evolution of Southern Alaska*, eds. K.D. Ridgway, J.M. Trop, J.M.G. Glen, and J.M. O'Neill, 345–377. Boulder, CO: Geological Society of America Special Paper 431.

Mark-Kurik, E. 2000. The Middle Devonian fishes of the Baltic States (Estonia, Latvia) and Belarus. *Courier Forschungs- Institut Senckenberg* 223:309–324.

Märss, T., M. Caldwell, P.Y. Gagnier, D. Goujet, P. Männik, T. Martma, and M. Wilson. 1998. Distribution of Silurian and Lower Devonian vertebrate microremains and conodonts in the Baillie-Hamilton and Cornwallis Island sections, Canadian Arctic. *Proceedings of the Estonian Academy of Sciences, Geology* 47:51–76.

Martin, G.C. 1905. The petroleum fields of the Pacific coast of Alaska, with an account of the Bering River coal deposits. *United States Geological Survey Bulletin* 250:1–64.

Martin, G.C. 1926. The Mesozoic stratigraphy of Alaska. *United States Geological Survey Bulletin* 776:1–493.

Martinez, R.N., P.C. Sereno, O.A. Alcober, C.E. Colomi, P.R. Renne, I.P. Montañez, and B.S. Currie. 2011. A basal dinosaur from the dawn of the dinosaur era in southwestern Pangaea. *Science* 331:206–210.

Merritt, R.D. 1985. Coal atlas of the Matanuska Valley. *Alaska Division of Geological and Geophysical Surveys Public-data File* 85–45:1–270.

Mertie, J.B., Jr. 1942. Tertiary deposits of the Eagle–Circle District Alaska. *United States Geological Survey Bulletin* 917-D:1–264.

Miller, E.L., A. Grantz, and S.L. Klemperer. 2002. *Tectonic Evolution of the Bering Shelf-Chukchi Sea-Arctic Margin and Adjacent Landmasses.* Boulder, CO: Geological Society of America Special Paper 360.

Molenaar, C.M. 1985. Subsurface correlations and depositional history of the Nanushuk Group and related strata, North Slope, Alaska. In *Geology of the Nanushuk Group and related rocks*, ed. A.C. Huffman, Jr., North Slope, Alaska: United States Geological Survey Bulletin 1614:37–59.

Molenaar, C.M., K.J. Bird, and A.R. Kirk. 1987. Cretaceous and Tertiary stratigraphy of northeastern Alaska. In *Alaskan North Slope Geology*, eds. I.L. Tailleur and P. Weimer, 513–528. Sacramento, CA: Society of Economic Paleontologists and Mineralogists, Pacific Section.

Moore, T.E., W.K. Wallace, K.J. Bird, S.M. Karl, C.G. Mull, and J.T. Dillon. 1994. Geology of northern Alaska. In *The Geology of Alaska. The Geology of North America G-1*, eds. G. Plafker and H.C. Berg, 49–140. Boulder, CO: Geological Society of America.

Mull, C.G. 1985. Cretaceous tectonics, depositional cycles, and the Nanushuk Group, Brooks Range and Arctic Slope, Alaska. *United States Geological Survey Bulletin* 1614:7–36.

Mull, C.G., D.W. Houseknecht, and K.J. Bird. 2003. Revised Cretaceous and Tertiary Stratigraphic Nomenclature in the Colville Basin, Northern Alaska. *United States Geological Survey Professional Paper* 1673. Version 1.0, http://pubs.usgs.gov/pp/p1673/index.html.

Miyaoka, R.T. 1990. Fossil locality map and fossil data for the southeastern Charley River Quadrangle, east-central Alaska: *United States Geological Survey*, Miscellaneous Field Studies Map, Map MF-2007, 1:100,000.

Nesbitt, S.J., P.M. Barrett, S. Werning, C.A. Sidor, and A.J. Charig. 2012. The oldest dinosaur? A Middle Triassic dinosauriform from Tanzania. *Biology Letters* 9:20120949.

Nokleberg, W.J., T.K. Bundtzen, R.A. Eremin, V.V. Ratkin, K.M. Dawson, V.I. Shpikerman, N.A. Goryachev et al. 2005. Metallogenesis and tectonics of the Russian far east, Alaska, and the Canadian cordillera. *United States Geological Survey Professional Paper* 1697:1–397.

Nokleberg, W.J., L.M. Parfenov, J.W.H. Monger, I.O. Norton, A.I. Khanchuk, D.B. Stone, C.R. Scotese, D.W. Scholl, and K. Fujita. 2000. Phanerozoic tectonic evolution of the circum-North Pacific. *United States Geological Survey Professional Paper* 1626:1–122.

Nokleberg, W.J. and D.H. Richter. 2007. Origin of narrow terranes and adjacent major terranes occurring along the Denali Fault and in the eastern and central Alaska Range, Alaska. In *Growth of a Collisional Continental Margin: Crustal Evolution of Southern Alaska*, eds. K.D. Ridgway, J.M. Trop, J.M.G. Glen, and J.M. O'Neill, 129–154. Boulder, CO: Geological Society of America Special Paper 431.

Ørvig, T. 1957. Notes on some Paleozoic lower vertebrates from Spitsbergen and North America. *Norsk Geologisk Tiddskrift* 37:285–353.

Ørvig, T. 1975. Description, with special reference to the dermal skeleton, of a new radotinid arthrodire from the Gedinnian of Arctic Canada. *Colloques Internationaux du Centre National de la Recherche Scientifique* 218:41–71.

Parrish J.T. and R.A. Spicer. 1988. Late Cretaceous terrestrial vegetation: A near-polar temperature curve. *Geology* 16:22–25.

Parrish, M.J., J.T. Parrish, J.H. Hutchinson, and R.A. Spicer. 1987. Late Cretaceous vertebrate fossils from the North Slope of Alaska and implications for dinosaur ecology. *Palaios* 2:377–389.

Pasch, A.D. and K.C. May. 1997. First occurrence of a hadrosaur (Dinosauria) from the Matanuska Formation (Turonian) in the Talkeetna Mountains of south-central Alaska. *Alaska Department of Natural Resources Professional Report* 118:99–109.

Pasch, A.D. and K.C. May. 2001. Taphonomy and paleoenvironment of a hadrosaur (Dinosauria) from the Matanuska Formation (Turonian) in south-central Alaska. In *Mesozoic Vertebrate Life*, eds. D.H. Tanke and K. Carpenter, 219–236. Bloomington, IN: Indiana University Press.

Pearce, N.J.G., J.A. Westgate, S.J. Preece, W.J. Eastwood, and P.T. Perkins. 2004. Identification of Aniakchak (Alaska) tephra in Greenland ice core challenges the 1645 bc date for Minoan eruption of Santorini. *Geochemistry, Geophysics, Geosystems* 5:Q03005, doi:10.1029/2003GC000672.

Perkins, P.L. 1971. The dipnoan fish *Dipterus* from the middle Devonian (Givetian) of Alaska. *Journal of Paleontology* 45:554–555.

Plafker, G. and H.C. Berg. 1994a. *The Geology of Alaska. The Geology of North America G-1*. Boulder, CO: Geological Society of America.

Plafker, G. and H.C. Berg. 1994b. Overview of the geology and tectonic evolution of Alaska. In *The Geology of Alaska. The Geology of North America G-1*, eds. G. Plafker and H.C. Berg, 989–1021. Boulder, CO: Geological Society of America.

Prindle, L.M. 1913. A geologic reconnaissance of the Circle quadrangle, Alaska. *United States Geological Survey Bulletin* 538:33–34.

Richter, D.H. 1976. Geologic map of the Nabesna quadrangle, Alaska. *United States Geological Survey Miscellaneous Geological Investigations Series* Map I-932, 1 sheet, scale 1:250,000.

Richter, D.H., C.C. Preller, K.A. Labay, and N.B. Shew. 2006. Geologic map of the Wrangell-Saint Elias National Park and Preserve, Alaska. *United States Geological Survey Scientific Investigations Series*, Map SIM-2877, 1:350,000.

Ridgway, K.D., J.M. Trop, J.M.G. Glen, and J.M. O'Neill, eds. 2007. *Growth of a Collisional Continental Margin: Crustal Evolution of Southern Alaska*. Boulder, CO: Geological Society of America Special Paper 431.

Ridgway, K.D., J.M. Trop, W.J. Nokleberg, C.M. Davidson, and K.R. Eastham. 2002. Mesozoic and Cenozoic tectonics of the eastern and central Alaska Range: Progressive basin development and deformation in a suture zone. *Geological Society of America Bulletin* 114:1480–1504.

Ridgway, K.D., J.M. Trop, and A.R. Sweet. 1997. Thrust-top basin formation along a suture zone, Cantwell basin, Alaska Range: Implications for the development of the Denali Fault system. *Geological Society of America Bulletin* 109:505–523.

Ringsmuth, K.J. 2007. Beyond the Moon Crater Myth: A new history of the Aniakchak landscape. *National Park Service, Research/Resources Management Report* AR/CRR-2207-63.

Rivals, F., M.C. Mihlbachler, N. Solounias, D. Mol, G.M. Semprebon, J. de Vos, and D.C. Kalthoff. 2010. Palaeoecology of the Mammoth Steppe fauna from the late Pleistocene of the North Sea and Alaska: Separating species preferences from geographic influence in paleoecological dental wear analysis. *Palaeogeography, Palaeoclimatology, Palaeoecology* 286:42–54.

Roehler, H.W. 1987. Depositional environments of the coal-bearing and associated formations of Cretaceous age in the National Petroleum Reserve in Alaska. *United States Geological Survey Bulletin* 1575:1–16.

Roehler, H.W. and G.D. Stricker. 1984. Dinosaur and wood fossils from the Cretaceous Corwin Formation in the National Petroleum Reserve, North Slope, Alaska. *Journal of the Alaska Geological Society* 4:35–41.

Rogers, R.R., C.C. Swisher, III, P.C. Sereno, A.M. Monetta, C.A. Forster, and R.N. Martinez. 1993. The Ischigualasto Tetrapod Assemblage (Late Triassic, Argentina) and $^{40}Ar/^{39}Ar$ dating of dinosaur origins. *Science* 260:794–797.

Salazar-Jaramillo, S., S.J. Fowell, P.J. McCarthy, J.A. Benowitz, M.G. Śliwiński, and C.S. Tomsich. 2016. Terrestrial isotopic evidence for a Middle-Maastrichtian warming event from the lower Cantwell Formation, Alaska. *Palaeogeography, Palaeoclimatology, Palaeoecology* 441:360–376.

Saltus, R.W. 2010. Matching magnetic trends and patterns across the Tintina Fault, Alaska and Canada: Evidence for offset of about 490 kilometers. *United States Geological Survey Scientific Investigations Report* 2007–5289-C:1–7.

Schrader, F.C. 1902. Geological section of the Rocky Mountains in northern Alaska. *Geological Society of America Bulletin* 13:233–252.

Shimer, G.T., P.J. McCarthy, and C.L. Hanks. 2014. Sedimentology, stratigraphy, and reservoir properties of an unconventional, shallow, frozen petroleum reservoir in the Cretaceous Nanushuk Formation at Umiat field, North Slope, Alaska. *American Association of Petroleum Geologists Bulletin* 98:631–661.

Skinner, M.F. and O.C. Kaisen. 1947. The fossil bison of Alaska and preliminary revision of the genus. *Bulletin of the American Museum of Natural History* 89:123–256.

Smith, W. 1925. Aniakchak Crater, Alaska Peninsula. *United States Geological Survey Professional Paper* 132-J:139–145.

Spicer, R.A. 2003. Changing climate and biota. In *The Cretaceous World*, ed. P. Skelton, 85–162. Cambridge: Cambridge University Press.

Spicer, R.A. and A.B. Herman. 2010. The Late Cretaceous environment of the Arctic: A quantitative reassessment based on plant fossils. *Palaeogeography, Palaeoclimatology, Palaeoecology* 295:423–442.

Spurr, J.E. 1900. A reconnaissance in southwestern Alaska in 1898. *United States Geological Survey 20th Annual Report* 7:31–264.

Stamatakos, J.A., J.M. Trop, and K.D. Ridgway. 2001. Late Cretaceous paleogeography of Wrangellia: Paleomagnetism of the MacColl Ridge Formation, southern Alaska, revisited. *Geology* 29:947–950.

Stensiö, E.A. 1927. The Downtonian and Devonian vertebrates of Spitsbergen. I, Family Cephalaspidae. *Skrifter om Svalbard og Ishavet* 12:1–391.

Stevens, C.H., V.I. Davydov, and D. Bradley. 1997. Permian Tethyan Fusilinina from the Kenai Peninsula, Alaska. *Journal of Paleontology* 71:985–994.

**60**
Alaska Dinosaurs

Tailleur, I. 1973. A skeleton in Triassic Rocks in the Brooks Range Foothills. *Arctic* 26:79–80.

Till, A.B., S.M. Roeske, D.C. Bradley, R. Friedman, and P.W. Layer. 2007. Early Tertiary transtension-related deformation and magmatism along the Tintina fault system, Alaska. In *Exhumation Associated with Continental Strike-Slip Fault Systems*, eds. A.B. Till, S.M. Roeske, J.C. Sample, and D.A. Foster, 233–264. Boulder, CO: Geological Society of America Special Paper 434.

Tomsich, C.S., P.J. McCarthy, A.R. Fiorillo, D.B. Stone, J.A. Benowitz, and P.B. O'Sullivan. 2014. New zircon U-Pb ages for the lower Cantwell Formation: Implications for the Late Cretaceous paleoecology and paleoenvironment of the lower Cantwell Formation near Sable Mountain, Denali National Park and Preserve, central Alaska Range, USA. In *ICAM VI: Proceedings of the International Conference on Arctic Margins VI*, Fairbanks, Alaska, May 2011, eds. D.B. Stone, G.K. Grikurov, J.G. Clough, G.N. Oakey, and D.K. Thurston, 19–60. St. Petersburg: VSEGEI, St. Petersburg.

Trop, J.M. 2008. Latest Cretaceous forearc basin development along an accretionary convergent margin: South-central Alaska. *Geological Society of America Bulletin* 120:207–224.

Trop, J.M. and K.D. Ridgway. 2007. Mesozoic and Cenozoic tectonic growth of southern Alaska: A sedimentary basin perspective. In *Tectonic Growth of a Collisional Continental Margin: Crustal Evolution of Southern Alaska*, eds. K.D. Ridgway, J.M. Trop, J.M.G. Glen, and J.M. O'Neill, 55–94. Boulder, CO: Geological Society of America Special Paper 431.

Trop, J.M., K.D. Ridgway, J.D. Manuszak, and P.W. Layer. 2002. Sedimentary basin development on the allochthonous Wrangellia composite terrane, Mesozoic Wrangell Mountains basin, Alaska: A long-term record of terrane migration and arc construction. *Geological Society of America Bulletin* 114:693–717.

Van Kooten, G.K., A.B. Watts, J. Coogan, V.S. Mount, R.F. Swenson, P.H. Daggett, J.G. Clough, C.T. Roberts, and S.C. Bergman. 1997. Geologic investigations of the Kandik area, Alaska and adjacent Yukon Territory, Canada. *Alaska Division of Geological and Geophysical Surveys Report of Investigation* 96-6A, p. 3 sheets, scale 1:200,000.

Vorobyeva, E.I. 1980. Observations on two rhipidistian fishes from the Upper Devonian of Lode, Latvia. *Zoological Journal of the Linnean Society* 70:191–201.

Wahrhaftig, C., S. Bartsch-Winkler, and G.D. Stricker. 1994. Coal in Alaska. In *The Geology of Alaska. The Geology of North America G-1*, eds. G. Plafker and H.C. Berg, 937–978. Boulder, CO: Geological Society of America.

Wilson, F.H., R.L. Detterman, and G.D. DuBois. 1999. Digital data for geologic framework of the Alaska Peninsula, southwest Alaska, and the Alaska Peninsula terrane. *United States Geological Survey Open-File Report OFR* 99–317.

Winkler, G.R. 2000. A geologic guide to Wrangell–St. Elias National Park and Preserve, Alaska: A tectonic collage of northbound terranes. *United States Geologic Survey Professional Paper* 1616:1–166.

Witte, K.W., D.B. Stone, and C.G. Mull. 1987. Paleomagnetism, paleobotany, and paleogeography of the Cretaceous, North Slope, Alaska. In *Alaskan North Slope Geology*, eds. I.L. Tailleur and P. Weimer, 571–579. Sacramento, CA: Society of Economic Paleontologists and Mineralogists, Pacific Section.

Wolfe, J.A. and C. Wahrhaftig. 1970. The Cantwell Formation of the central Alaska Range. *United States Geological Survey Bulletin* 1294-A:41–46.

Zazula, G.D., D.G. Froese, S.A. Elias, S. Kuzmina, and R.W. Mathewes. 2007. Arctic ground squirrels of the mammoth-steppe: Paleoecology of Late Pleistocene middens (~24000–29450 14 C yr BP), Yukon Territory, Canada. *Quaternary Science Reviews* 26:979–1003.

# 4 The Bones

## INTRODUCTION

If we define the Arctic region as that area of the Earth north of the Arctic Circle, then the Arctic encompasses approximately 510,000,000 square kilometers, or approximately 6% of the Earth's surface. Of course, that percentage includes the Arctic Ocean, which is approximately 15,500,000 square kilometers. Subtracting the Arctic Ocean from the total area of the Arctic still leaves a substantial amount of real estate. Within this vast region, there is only a smattering of dinosaur fossil records, non-avian or avian, outside of Alaska, and those records are predominantly found in Canada, Norway, and Russia (e.g., Lapparent, 1962; Rouse and Srivastava, 1972; Edwards et al., 1978; Hurum et al., 2006; Godefroit et al., 2009; Evans et al., 2012). If we refer back to the Victorian concept of saurians, which included marine reptiles and pterosaurs as well as dinosaurs, the record is only somewhat expanded. If we include fish and mammals, the record is expanded a little more. Figure 4.1 is a generalized cladogram showing the evolutionary relationships between many of the animals discussed in this chapter. Within this entire suite known Alaskan fossil vertebrates, dinosaurs are the best represented of Mesozoic vertebrates and tell an important story about biodiversity and adaptation in an extreme environment.

Therefore, with respect to this expanded perspective on the Mesozoic vertebrate fossil record of the high latitudes, the marine reptile record within Alaska is thus far restricted to the first half of the Mesozoic. While marine reptiles such as thalattosaurs and plesiosaurs are known (Weems and Blodgett, 1996; Adams, 2009), ichthyosaurs are the most common marine reptiles reported across the state (Tailleur et al., 1973; Parrish et al., 2001; Adams, 2009; Druckenmiller and Maxwell, 2014; Druckenmiller et al., 2014). These various reports of ichthyosaurs span from the North Slope to the southeastern part of modern Alaska (Figure 4.2). In the southeastern part of the state, ichthyosaurs tended to dominate the open sea, while the thalattosaurs dominated the near-shore coastal environments (Adams, 2009). The Triassic contains the more robust record when compared to the Jurassic, and there are no Cretaceous records yet known for marine reptiles. The Triassic ichthyosaurs, as members of the Shastasauridae, tend to be fairly large animals (Figure 4.3; Adams, 2009; Druckenmiller et al., 2014). In addition to these reptiles there have been reports of the lobe-finned fish, the coelacanth, and the now extinct ray-finned paleoniscoid fishes from the Triassic Shublik Formation of northern Alaska (Patton and Tailleur, 1964; Zheleznov and Okuneva, 1972). The rocks containing these fossils were deposited far to the south of their current latitudes (Lawver et al., 2002) during the Triassic, and thus do not contribute to the discussion of ancient Arctic ecosystems.

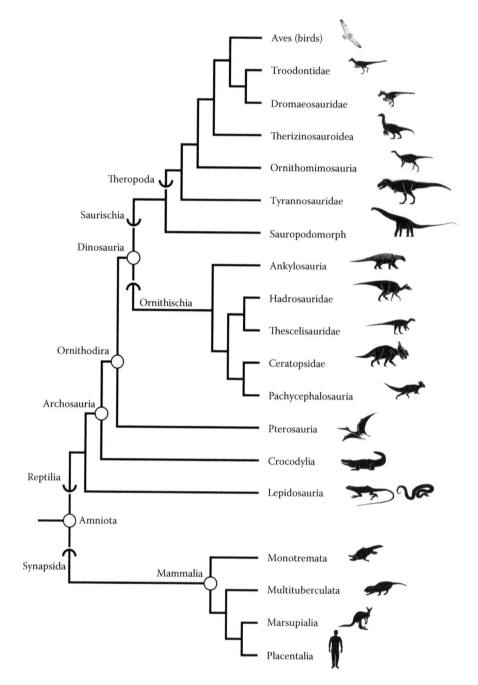

**FIGURE 4.1** Generalized cladogram showing the evolutionary relationships of the primary animals discussed in this chapter.

**FIGURE 4.2** Example of Triassic rock outcrop in Alaska. These exposures of rocks are found in Wrangell–St. Elias National Park in southeastern Alaska.

As discussed in Chapter 3, by about 100 million years ago during the Cretaceous Period the tectonic assembling of terranes was approaching the appearance of modern Alaska, and throughout the Cretaceous the diversity of species of all vertebrates represented in the fossil record of Alaska is sparse. Older Cretaceous dinosaur bones are now reported in a few places within Alaska from rocks deposited at higher latitudes, such as the Matanuska Formation and the Nanushuk Formation, and those dinosaur remains are primarily attributed to members of the Ornithopoda (Figure 4.4; Pasch and May, 1997; Fiorillo, 2006). Younger Cretaceous rocks in Alaska contain a much better record, and by far the richest dinosaur fauna has been recovered from the Prince Creek Formation of northern Alaska (Fiorillo and Gangloff, 2000, 2001; Rich et al., 2002; Gangloff et al., 2005; Fiorillo, 2008; Fiorillo et al., 2009; Gangloff and Fiorillo, 2010; Fiorillo and Tykoski, 2012, 2013, 2014; Tykoski and Fiorillo, 2013). With that said, however, the faunal list of latest Cretaceous dinosaurs from all of Alaska—five taxa of theropods and five taxa of ornthischians known from skeletal remains (Table 4.1)—is low when compared to comparably aged dinosaur faunas found in more southerly latitudes.

Since the mid-1980s, significant effort has been spent collecting dinosaurian material in northern Alaska, and while this window of time of study does not compare in

**FIGURE 4.3**   Vertebrae attributed to *Shonisaurus* from Triassic exposures on Hound Island, Alaska. (Photograph courtesy of Thomas Adams, Witte Museum.)

length to exploration and collecting in southern regions, it is long enough to suggest that a reduction in taxonomic diversity in the ancient higher latitudes may be real.

In the modern world, taxonomic diversity generally decreases in a gradient from lower equatorial latitudes to higher-polar latitudes (e.g., Bee and Hall, 1956; Pianka, 1966; Pisanty-Baruch et al., 1999; Hawkins et al., 2003). This diversity change is linked to the climatic variance between low and high latitudes and resource availability correlated with climate. Even though the climate of the Cretaceous was more equable—that is, less extreme—than that of the modern world, climate models of the Late Cretaceous show a pronounced temperature gradient from low to high latitudes (DeConto et al., 1999).

How modern climate influences modern species richness and biodiversity has long been a question (e.g., von Humboldt, 1808; Schall and Pianka, 1977, 1978;

5 cm

**FIGURE 4.4**   Distal humerus of an ornithopod dinosaur from the Nanushuk Formation. The specimen was recovered along the Kaolak River by Don Triplehorn (Emeritus, University of Alaska). (University of Alaska Museum of the North specimen AK-280-V-02.)

Huston, 1979; Wright, 1983; Currie, 1991; Mittleback et al., 2001; Hawkins et al., 2003; Evans et al., 2005; Clarke and Gaston, 2006). While it is widely recognized that energy and water drive plant diversity in high latitudes (e.g., Jefferies et al., 1992; Walker et al., 1998; McKane et al., 2002), it is less clear what aspects impact modern animals. Given this uncertainty, for now it will suffice that though the exact climatic mechanisms that influence biota are not fully understood for extant organisms, the pattern of apparent low-taxonomic diversity of the Alaskan dinosaur fauna compared to the more southerly latitudes of North America during the Late Cretaceous of North America does follow the same biogeographic phenomenon as observed in the modern world. Here we look at the primary dinosaurs and other fossil vertebrates known from Alaska to better understand the vertebrate biodiversity of the ancient Alaskan Arctic.

## THE THEROPOD DINOSAURS

Most of the theropod dinosaur remains from the Cretaceous of Alaska are isolated teeth. Currie et al. (1990) showed that the teeth of small theropods provide reliable criteria for generic and even species identification. Subsequent workers followed Currie and others (1990) to document theropod faunas from a variety of localities

## TABLE 4.1
## Dinosaurs Found in the Latest Cretaceous of Alaska

Saurischia
    Dromaeosauridae
        *Dromaeosaurus*
        *Saurornitholestes*
    Troodontidae
        *Troodon*
    Ornithomimisauria
    Tyrannosauridae
        *Nanuqsaurus*
Ornithischia
    Thescelosauridae
        cf. *Parksosaurus*
    Hadrosauridae
        *Edmontosaurus*
    Nodosauridae
        *Edmontonia*
    Pachcephalosauridae
        *Alaskacephale*
    Ceratopsidae
        *Pachyrhinosaurus*

*Note:* See text for detailed discussion of these taxa.

around the world (Rowe et al., 1992; Chure, 1994; Fiorillo and Currie, 1994; Zinke and Rauhut, 1994; Fiorillo and Gangloff, 2000; Sankey, 2001; Sankey et al., 2005), employing isolated teeth and the information they provide. Though there is some evidence of heterodonty, or different tooth shapes along the jaw, in some theropods, theropod teeth are generally laterally compressed, curved, with serrations on the anterior and posterior ridges, or carinae (Holtz et al., 2004; Tykoski and Rowe, 2004). Such morphology is like most carnivorous lizard teeth, allowing some workers to model theropod functional food use patterns using the teeth of lizards (Abler, 1992).

## DROMAEOSAURUS

*Dromaeosaurus* is a small theropod, perhaps 2 meters long, found in Cretaceous rocks in western North America, and is famous for its enlarged second toe wielding a sharply curved "sickle" claw. This specialization of the toe aside, the features of the skeleton show *Dromaeosaurus* to be among the least specialized small theropods of the time (Currie et al., 1990; Currie, 1995). *Dromaeosaurus* is also the namesake for the group, the Dromaeosauridae.

Originally found in the Red Deer River badlands of Alberta, this dinosaur has achieved a certain level of fame, yet remains surprisingly poorly understood when compared to other dinosaurs. The teeth resemble teeth of most other theropods, but in this dinosaur they are strongly laterally compressed such that the fore–aft measurement can be almost twice the width. Denticles, or serrations, line both the anterior and posterior margins of the tooth, and are broad and chisel-like. These teeth also commonly have shallow and often poorly defined blood grooves that extend onto the surface near the base of the tooth. The morphology of these teeth was well adapted to slicing and processing the meat from the bones of their prey.

*Dromaeosaurus* is known from the Prince Creek Formation from a handful of isolated teeth (Fiorillo and Gangloff, 2000). There is no reliable way to discern if the teeth are from the upper jaw or the lower jaw, so determining how many individuals are represented is not clear. Based on the relatively robust skull known from skeletons found elsewhere, and the presence of large teeth, there is no question *Dromaeosaurus* was a powerful predator. However, the second toe aside, given that this animal does not display additional skeletal specializations that might have implications for its paleobiology, these dinosaurs likely then were ecological generalists—that is, they were not restricted to specific niche spaces, and instead made use of a variety of food resources. So the ecological generalist lifestyle was well suited to existence throughout western North America in the later Cretaceous, but being a generalist does not seem to have presented a particular advantage within the predator guild of the ancient north.

### SAURORNITHOLESTES

Like *Dromaeosaurus*, *Saurornitholestes* was a small, bipedal predatory dinosaur known from the Cretaceous of western North America, including Alaska, and is also a member of the dromaeosaurids. Also like *Dromaeosaurus*, *Saurornitholestes* had a sickle-like claw on each of the inner, or more technically the second, toes of its feet. *Saurornitholestes* was first found in Alberta but in contrast to *Dromaeosaurus*, it is now known from several very good specimens (e.g., Colbert and Russell, 1969; Currie, 1995). In the lower latitudes, this is the most common small predatory dinosaur found in the Late Cretaceous rocks of Alberta and Montana (Fiorillo and Gangloff, 2000).

The denticles of the teeth of *Saurornitholestes* differ from those of *Dromaeosaurus* in that those of the former are minute or absent on the anterior edge of the teeth compared to those of *Dromaeosaurus* (Currie et al., 1990). Further, the denticles on the posterior edge of the teeth of *Saurornitholestes* are more pronounced and pointed than the denticles on the teeth of *Dromaeosaurus* (Currie et al., 1990). Unlike these more southern latitude geographic regions, *Saurornitholestes* is an uncommon to rare component of the small predatory guild of dinosaurs in Alaska (Fiorillo and Gangloff, 2000). The dominance of this dinosaur in Montana and Alberta locations suggests some level of competitive advantage for this small theropod in the more southern latitudes, an advantage that seems to not have been in place in the higher latitudes of ancient Alaska.

### TROODON

*Troodon* is a member of the Troodontidae, a group that, along with the Dromaeosauridae, share many anatomical features with modern birds, such as longer forelimbs, modifications for greater wrist movements, and modifications of the pelvis that shift the center of gravity. Members of the family are small- to medium-sized predators known for, among other things, articular facets on the second toe that presumably gave the ability to hyperextend this digit, creating a sickle-like motion for the enlarged claw as it sliced into potential prey. Though enlarged, the claw on this second digit is not anywhere near as hypertrophied as in dromaeosaurs, even if the second digit modifications seem to have been similarly lethal. These differences in claws suggest that perhaps *Troodon* was the more nimble and fast hunter, while dromaeosaurs were the more powerful hunters.

Troodontids have large brains in relation to their body size when compared to other non-avian dinosaurs (Russell and Seguin, 1982), a proportion that is comparable to some living birds, specifically members of the palaeognaths, or flightless birds, which have the lowest encephalization quotients among birds (Corfield et al., 2008). Also unusual among theropods is the structure of the ear region of troodontids. These dinosaurs had enlarged middle ear cavities, suggesting that troodontids had a more acute sense of hearing compared to other theropods (Currie, 1985).

As one of the first dinosaurs described from North America (Leidy, 1856), the history of *Troodon* is long, though over much of that history the taxon was poorly understood. Taxonomic confusion stemmed from a poor fossil record. Isolated teeth found by Ferdinand Vandeveer Hayden while conducting a geological reconnaissance in the mid-1850s in central Montana were later described by Joseph Leidy (1856). Leidy erected the genus *Troodon*, but he considered *Troodon* a lizard rather than a dinosaur. Later workers attributed this animal to a variety of theropod and ornithischian groups (e.g., Nopsca, 1901; Gilmore, 1924). While the family Troodontidae was originally defined by Charles Gilmore in 1924, he considered *Troodon* to be an ornithischian dinosaur. Loris Russell (1948) disagreed with this assessment, and reassigned the Troodontidae to theropods based on his detailed description of a jaw of *Troodon*. In his analysis of troodontid specimens, including a then-new discovery of a *Troodon* jaw, Philip Currie (1987) established unequivocally the identity of Troodontidae as theropod dinosaurs.

In contrast to other small theropods from the Prince Creek Formation, *Troodon* is known from both isolated teeth and non-dental cranial material (Fiorillo and Gangloff, 2000; Fiorillo, 2008; Fiorillo et al., 2009). Based on the frequency of isolated teeth from this rock unit, *Troodon* is the most common theropod of the ancient Arctic (Fiorillo and Gangloff, 2000; Fiorillo, 2008). The fortuitous discovery of two partial braincases attributable to *Troodon formosus* confirmed presence of this animal in northern Alaska (Fiorillo et al., 2009).

*Troodon formosus* teeth in Alaska are twice the size of those found in the southern latitudes (Fiorillo, 2008). Broadly related vertebrate animals with similarly shaped teeth should exhibit similar predictable relationships between body size and tooth size, given that such animals tend to have similar diets (Fortelius, 1990). Thus, given the similarity in shape of some lizard teeth to theropod teeth, it is relevant that some lizards show a positive correlation between tooth size and body size (Mateo and Lopez-Jurado, 1997; Townsend et al., 1999). Drawing from this relationship with modern lizards, the *Troodon* teeth reflect a larger body size for the Arctic population of *Troodon*.

In the mid-19th century, Carl Bergmann proposed this zoological rule that stated, within a broadly distributed genus, species found in colder environments were larger than species found in warmer environments. In the 1950s, it was argued that the rule was flawed and that many vertebrates show a body size independent of latitude (Scholander, 1955, 1956; McNab, 1971; Irving, 1972; Geist, 1987). More recently, however, Ashton (2002a) suggested that body size correlated with latitude in amphibians, but a similar look at Bergmann's rule as applied to turtles showed no such correlation (Angielczyk et al., 2015). In a separate study, Ashton (2002b) made a significantly stronger case for the application of Bergmann's rule to birds. Underlying Bergmann's rule is the observation that body volume increases at a rate of a linear size measurement times $10^3$ while surface area increases at a rate of $10^2$, so that larger individuals have a relatively smaller surface area from which to lose body heat, which would provide an advantage to larger animals in colder northern climes in many cases.

A notable exception to Bergmann's rule is shown in grizzly bears (*Ursus arctos*). These bears once had a latitudinal range extending from the modern Arctic to central Mexico (Servheen, 1984). Their current range is much reduced, though still substantial (Kays and Wilson, 2002). The latitudinal distribution in body size of these bears is shown in Table 4.2. The body mass of adult male grizzly bears in Yellowstone National Park, Wyoming, is approximately 190 kg (Hilderbrand et al., 1999), and somewhat higher in Denali National Park (Hilderbrand et al., 1999), while farther north in the western Brooks Range the body mass diminishes to approximately the body mass of bears in Yellowstone National Park (Hilderbrand et al., 1999). Most noticeable along this latitudinal transect is the enormous increase in mass of the

---

**TABLE 4.2**

**The Average Mass of Adult Male Grizzly Bears from Selected Localities from Wyoming to Northern Alaska**

| Location | Approximate Latitude | Weight (kg) |
|---|---|---|
| Western Brooks Range, Alaska | 68°N | 190 |
| Denali National Park, Alaska Range | 63°N | 220 |
| Kenai National Wildlife Refuge, Alaska | 60°N | 380 |
| Katmai National Park, Southwest Alaska | 58°N | 375 |
| Yellowstone National Park, Wyoming | 44°N | 190 |

*Source:* Figure 4.1 in Hilderbrand, G.V. et al. 1999. *Canadian Journal of Zoology* 77: 132–138 and Blanchard, B.M. 1987. Size and growth patterns of the Yellowstone grizzly bear. In *Bears—Their Biology and Management: Proceedings of the Seventh International Conference on Bear Research and Management*, Williamsburg, VA, ed. P. Zager, 99–107. February 21–26, 1986, and Plitvice Lakes, Yugoslavia, March 2–5, 1986. Washington, DC: International Association for Bear Research and Management.

*Note:* The localities are listed from north to south. The proxy latitude for Western Brooks Range in this table is for Noatak National Preserve. Note that the weight of these bears does not follow a latitudinal gradient.

---

bears in coastal south-central and southwestern Alaska (Hilderbrand et al., 1999), and the reversal of the trend in the Brooks Range.

Given the general nature of Bergmann's rule in the modern world, it would be highly speculative to invoke the rule to account for the larger size of the northern form of *Troodon* compared to its southern counterparts. Alternate models may account for the occurrence of large teeth of *Troodon* in the high latitudes, even though the hypothesis of Bergmann's rule cannot be rejected.

Internal processes, such as habitat productivity, and external processes, such as emigration and immigration, affect the number of species in a modern community (Schluter and Ricklefs, 1993). Constituent species in modern ecosystems partition resources within their respective environments (Mackie, 1970; Pianka et al., 1979; Freeman, 1981; Thomas and Edmonds, 1984; Grant, 1986), because no two sympatric species share exactly the same niche (Hutchinson, 1978; Schluter and Ricklefs, 1993).

Many of the same theropod species recur through several rock units in the Late Cretaceous, and as a result have been interpreted as being sympatric. There are at least four morphologically similar theropod taxa within latest Cretaceous ecosystems, or more specifically Campanian-Maastrichtian stages, of western North America: *Dromaeosaurus, Saurornitholestes, Troodon,* and *Richardoestesia* (Currie et al., 1990; Rowe et al., 1992; Fiorillo and Currie, 1994; Fiorillo, 1998a; Fiorillo and Gangloff, 2000; Sankey, 2001; Sankey et al., 2005). This composition is intriguing because nowhere in modern ecosystems do four morphologically similar carnivorous vertebrates coexist in one ecosystem.

Rather, in modern ecosystems, such as the canids of the Greater Yellowstone ecosystem (Clark et al., 1999) and the jackals in East Africa (Van Valkenburgh and Wayne, 1994), three morphologically similar carnivore systems are present. Several factors (e.g., Johnson et al., 1996; Crabtree and Sheldon, 1999) have been identified regarding the sympatry of canids in ecosystems (Table 4.3). In the East African ecosystem and the Greater Yellowstone ecosystem, shape divergence (e.g., Van

---

**TABLE 4.3**

**Factors Affecting Canid Sympatry in Modern Ecosystems**

Character Displacement
Home-Range Interspersion
Interference Competition
Scavenging
Size Dominance
Spatial and Temporal Partitioning

*Source:*  Adapted from Crabtree, R.L. and J.W. Sheldon. 1999. Coyotes and canid coexistence in Yellowstone. In *Carnivores in Ecosystems: The Yellowstone Experience*, eds. T.W. Clark et al. Kareiva, 127–163. New Haven, CT: Yale University Press.

*Note:*  Similar factors might also have influenced sympatry among morphologically similar theropods in Cretaceous ecosystems.

Valkenburgh and Wayne, 1994) or size divergence (Crabtree and Sheldon, 1999) among these carnivore taxa compress niche spaces. The sympatry among the East African canids might relate to the abundant diversity and biomass available (Crabtree and Sheldon, 1999). In contrast, Van Valkenburgh and Molnar (2002) suggested that theropods had reduced metabolic rates compared to mammals, accounting for reduced trophic diversity in theropod communities and allowing more sympatric carnivorous taxa.

To accommodate increased theropod diversity within a single community, additional ecological models have been suggested based on *Troodon*. Holtz et al. (1998) suggested that *Troodon* may have been omnivorous based on tooth denticle morphology, thereby utilizing the niche space that differed from the other theropods. Temporal or spatial partitioning, rather than diet separation, has also been suggested as a mechanism for *Troodon* to fit into the predator guild of the Cretaceous ecosystem of western North America (Russell and Seguin, 1982; Fiorillo and Gangloff, 2000). *Troodon* has large eyes, suggesting it may have been most active in low-light conditions during dawn or dusk within temperate climates (Russell and Seguin, 1982), or in the dominant low-light conditions of ancient high latitudes (Fiorillo and Gangloff, 2000; Fiorillo, 2008), a situation somewhat analogous to the partitioning of niche space by time of hunting that is observed in some modern Marsh Hawks and Short-eared Owls (Figure 4.5). This adaptation by *Troodon* for low-light conditions (sensu Russell and Seguin, 1982) likely explains why nearly 75% of the isolated theropod teeth found in the Prince Creek Formation and be attributed to *Troodon*, that this dinosaur thrived within an ecosystem where the dominant physical aspect of the environment was low-angle light and an annual light input less than that in the lower latitudes (Spicer, 2003).

Wear patterns on teeth provide a qualitative measure of the texture of the food consumed by particular taxa (Walker et al., 1978; Kay and Covert, 1983; Teaford and Walker, 1984; Teaford, 1985; Taylor and Hannam, 1987; Walker and Teaford, 1989; Ungar, P.S., 1994; Hotton et al., 1997; Fiorillo, 1998b; Ungar, 1998; Rivals and Deniaux, 2003; Goswami et al., 2005). Hard food items such as bone, gritty foods, invertebrate shells, and carapaces yield heavy scratches and pits on teeth, whereas softer food items yield patterns of fine wear. Similar patterns of microwear between populations of *Troodon* teeth in Alaska and Montana indicate that these animals consumed similar food items (Fiorillo, 2008). Evidently, *Troodon* ate foods of similar abrasiveness regardless of latitude.

Ecological controls could account for the statistically larger northern forms of *Troodon* using modern analogs to account for this size difference. Van Valkenburgh and Molnar (2002) compared mammalian predators with dinosaurian predators and stated that sympatric large carnivores typically differ along any of three lines: body size, dental morphology, or skeletal anatomy. Their study also showed that the cranial proportions of theropods parallel those of canids more so than any other mammalian predator group. Given the greater estimated body size of northern individuals of *Troodon* based on tooth dimensions, parallels from observations of canid ecology might be drawn to *Troodon*.

When a specific genotype produces different phenotypes in response to distinct environmental conditions, that phenomenon is termed *phenotypic plasticity*

**FIGURE 4.5** Depiction of the niche separation relationship between a Northern Harrier (*Circus cyaneus*) and a Short-eared Owl (*Asio flammeus*). Though both feed on rodents, the Northern Harrier is active during the daytime (a) and the Short-eared Owl is active during the evenings and mornings (b), thereby reducing direct competition for resources. This illustration is based on the author's birding observations near the Dulles Airport in Virginia in the early 1980s.

(Pigliucci, 2001). A number of researchers have examined the plasticity of body size in carnivores such as coyotes (*Canis latrans*) and grizzly bears (*Ursus arctos*). The observed larger body size of coyotes in their eastward range expansion has been hypothesized to result from phenotypic response or genotypic selection arising from changes in prey size and prey abundance (Thurber and Peterson, 1991). In another canid species, the evolution of size differences among Boreal, Algonquin, and Tweed wolves (*Canis lupus*) of Ontario has been suggested to be a response to prey size (Schmitz and Lavigne, 1987), whereas greater food supply has been offered as a cause by others (Rosenzweig, 1968; Thurber and Peterson, 1991). Similarly, coyote body size differences may have been the result of similar phenotypic response to enhanced food supply, possibly involving no genetic selection (Thurber and Peterson, 1991). A similar pattern of body size being influenced by food and resource abundance may be the mechanism behind the larger size in northern forms of *Troodon* (Fiorillo, 2008) rather than attributing the pattern to Bergmann's Rule.

## ORNITHOMIMISAURIA

Commonly referred to as the "ostrich dinosaurs," ornithomimids are known from both North America and Asia (Kobayashi and Lü, 2003; Lee et al., 2014; McFeeters et al., 2016, 2017) and are Cretaceous in age. This unusual group of theropods had long necks and long limbs, and while early forms of this group did have teeth, the latest Cretaceous species were edentulous—that is, they lacked teeth. At least one ornithomimid species, *Sinornithomimus dongi*, was likely to have been gregarious (Kobayashi and Lü, 2003). The bonebeds containing *Sinornithomimus dongi* also contained many juveniles. Because it seems the adults were better adapted for faster running than the juveniles, and *Sinornithomimus dongi* was likely herbivorous (Kobayashi et al., 1999), it may be that they herded to protect the juveniles, similar to behavior seen in many modern herbivorous of mammals.

A metatarsal foot bone attributed to the more advanced ornithomimosaurs of the family Ornithomimidae was found while excavating the Kikak-Tegoseak Quarry (Gangloff, personal communication 2001), yet it remains undescribed. A second specimen, a partial metatarsal, was recovered from a different locality, Old Bone Beach, located several miles downriver from the Kikak-Tegoseak Quarry. This second specimen is referred to the Ornithomimisauria (Watanabe et al., 2013). This specimen resembles the fourth metatarsal of both ornithomimosaurs and juvenile tyrannosaurs. Given that at the time of their study dwarfed tyrannosaurs were not known from the ancient Arctic, Watanabe et al. (2013) referred the specimen to the former group. More recent work has shown that a dwarfed tyrannosaur roamed ancient Alaska (Fiorillo and Tykoski, 2014; Fiorillo et al., 2016); thus it may be that we have to wait until more definitive material is found before placing ornithomimosaurs within the Late Cretaceous terrestrial ecosystem of Alaska, although they would be expected, as they are found on both sides of the Bering Strait, in Asia, and further south in North America.

### NANUQSAURUS

*Nanuqsaurus* is a diminutive tyrannosaur from the Prince Creek Formation of northern Alaska (Fiorillo and Tykoski, 2014). Tyrannosaurs are typically viewed as being very large theropods. The Tyrannosauridae contains two subfamilies, the Albertosaurinae, defined as all tyrannosaurids closer to *Albertosaurus sarcophagus* than to *Tyrannosaurus rex*, and the more taxonomically diverse Tyrannosaurinae, defined as all tyrannosaurids closer to *T. rex* than to *Albertosaurus sarcaophagus*. While both groups represent very large, bipedal predatory dinosaurs, the members of the Albertosaurinae tend to have a body length about three-quarters of that observed in *T. rex* and a reconstructed weight of only about half that of *T. rex*. The former subfamily is found in North America, while the Tyrannosaurine is found in both Asia and North America. As big toothy animals, tyrannosaurs have arguably the most public visibility of any type of dinosaur, filling a role similar in popularity to an all-star centerfielder of the New York Yankees. Almost immediately after their discovery, tyrannosaurs captivated the public and the scientific community. Shortly after the first descriptions of these animals over a century ago (Osborn, 1905), *T. rex*

**FIGURE 4.6**   Screenshots from *Superman: The Arctic Giant* (1942, Fleischer Studios).

made its screen debut in 1919, appearing in *The Ghost of Slumber Mountain*, a box office smash. A couple of decades later, Paramount Pictures united this cultural icon with another in the animated short film *Superman: The Arctic Giant* (1942; Figure 4.6).

Like any other group of dinosaurs, the study of these meat-eating dinosaurs is fascinating. Unlike other groups of dinosaurs, however, the study of theropod dinosaurs, and in particular the study of that most iconic of dinosaurs, *T. rex* and its close kin, can sometimes also include a significant amount of emotional attachment. Take for example the debate that occasionally surfaces as to whether or not *T. rex* was a predator chasing down live prey or a scavenger (see Brusatte, 2012; Carpenter, 2013 for summaries). The emotional attachment to the big, fierce, "king-of-beasts" interpretation was shown brilliantly by Bill Watterson in his Calvin and Hobbes comic strip in 1993. Calvin, a precocious 6-year-old, had taken on the debate about tyrannosaurs as "fearsome predators or disgusting scavengers?" as a school project. After several days of comic strips where Calvin weighed the merits of both sides of the debate, he had to read his paper in front of the class (Figure 4.7). Calvin's terse conclusion was that tyrannosaurs were clearly predators chasing down live prey, because "it would be bogus if they ate things that were already dead."

Given the passion for tyrannosaurs, the past several years have witnessed an abundance of new specimen discoveries and subsequent research on the group. The number of recognized tyrannosaur species has increased and presented challenges to older assumptions about the adaptability and evolution of these predatory dinosaurs (Brusatte et al., 2010). However, for all the attention given these animals, most of what we know about them has been gathered from fossils collected at sites in the low- to mid-latitudes of North America and Asia.

**FIGURE 4.7**   Calvin and Hobbes©. (Bill Watterson, 1993. Reprinted with permission of Andrews McMeel Syndication. All rights reserved.)

In the Arctic, after years of study of some fragmentary remains of a predatory dinosaur from the Prince Creek Formation, and while others were building the database on what we think we know about tyrannosaurs, my colleague Dr. Ron Tykoski and I started to think we had a new type of tyrannosaur. What made our study so interesting to us was that not only did it seem to be a new tyrannosaur, but it was only about half the size of its close cousin *T. rex*. When we presented our arguments at an annual meeting of the Society of Vertebrate Paleontology, we had clearly polarized the audience, as the responses were either along the lines of how intriguing the data were, or that my colleague and I were complete buffoons. The scientific method welcomes debate as ideas are rigorously tested and retested, but even so, the visceral response to our talk by some members of the audience was interesting, to say the least.

Previous reports of tyrannosaurs from northern Alaska had been based on a small number of isolated teeth from the bonebeds found in the bluffs of the Prince Creek Formation along the Colville River (Gangloff, 1994, 1998; Fiorillo and Gangloff, 2000). In the absence of non-dental skeletal data for comparisons, these isolated teeth had been attributed to southern forms of tyrannosaurids (Gangloff, 1994, 1998). The broader tyrannosaur determination was justified by a couple of features exhibited on these teeth. One is that some of the teeth had a basal D-shaped cross-section and lack of anterior and posterior carinae, or ridges, which is characteristic of teeth from the premaxilla of tyrannosaurids (Fiorillo and Gangloff, 2000). Another feature that separated these teeth from other groups of theropods is the shape of the denticles that run along the carinae. These denticles are similar to those of dromaeo-saurs but are distinguished from the teeth of these dinosaurs by the diagnostic blood groove between the denticles that extends obliquely toward the base of the tooth (Currie et al., 1990; Abler, 1992).

In 2006, parts of three closely associated loose blocks were collected from the Kikak-Tegoseak Quarry operation along the Colville River. The blocks were prepared and resulted in the recovery of parts of a tyrannosaur skull. The fragments—a piece of the ascending ramus of the right maxilla, a partial skull roof including parts of both frontals, parietals, and right laterosphenoid, and a section of the front end of the left dentary through the first nine tooth positions and part of the tenth—proved to be enlightening (Figure 4.8).

Though there seemed to be only a limited amount of fossil material representing the entire skull, enough information was preserved to test the relationships of the taxon within the framework of two recent cladistic analyses of tyrannosauroid phy-logeny (Brusatte et al., 2010; Loewen et al., 2013). This Alaskan tyrannosaurid skull showed the presence of a thin, rostrally forked, median spur of the fused parietals on the top of the skull roof that overlaps and separates the frontals within the sagittal crest. The skull also had a frontal with a long process pointed rostrally that separates the prefrontal and lacrimal facets, and the two dentary teeth/alveoli are much smaller than the dentary teeth/alveoli that follow.

Inputting the anatomical data from the Alaskan material into a phylogenetic analysis, it was clear that this Arctic theropod was a derived tyrannosaurine, closely related to the group that includes *Tyrannosaurus* and the Asian tyrannosaur *Tarbosaurus*. However, the few pieces also possessed unique anatomical features

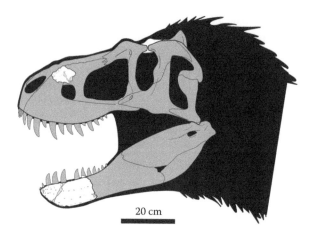

20 cm

**FIGURE 4.8** Drawing of the skull of the Arctic tyrannosaur, *Nanuqsaurus hoglundi*, from the Prince Creek Formation. The bones recovered for this animal are shown in white.

that justified the erecting of a new taxon, *Nanuqsaurus* (Fiorillo and Tykoski, 2014). Given its far northern life as the presumed top predator within the Cretaceous ecosystem, the name *Nanuqsaurus* means "polar bear lizard" (Figure 4.9; Fiorillo and Tykoski, 2014). Further statistical analyses showed that the differences between *Nanuqsaurus* and other derived tyrannosaurines were comparable to the differences for some of the more basal tyrannosauroids such as *Appalachiosaurus* and *Dryptosaurus*, where both of these latter dinosaurs are known from more complete material (Fiorillo and Tyksoski, 2014). So, for *Nanuqsaurus* the lesson was that it

**FIGURE 4.9** Reconstruction of *Nanuqsaurus hoglundi* in the Cretaceous environment in northern Alaska. (Artwork by Karen Carr.)

was not necessarily the amount of skeletal material that is important so much as how informative the material recovered is.

Many tyrannosaurs are of a body size near the upper limits of theropods. Thus, the estimated skull length of 600–700 mm in a mature individual of *Nanuqsaurus* suggests that this animal was smaller than contemporary related tyrannosaurids from the lower latitudes (Fiorillo and Tykoski, 2014), a bit ironic given that the 1942 Superman cartoon references the northern tyrannosaur as an "Arctic Giant." Typical large-bodied tyrannosaurs are also recognized as having enhanced olfaction, and their large body size suggests the animals were active hunters, rather than scavengers (Carbone et al., 2011). Although *Nanuqsaurus* was diminutive with respect to other tyrannosaurines, the skull shows evidence that it too had an enhanced sense of smell.

With respect to modern mammalian carnivores, the representatives at the upper size limits are constrained by food intake (Carbone et al., 2007). In the profoundly seasonal Cretaceous Arctic environment, like today's Arctic, the widely varying light regime affected biological productivity. Therefore food resource availability was likely seasonally variable, and so the smaller body size of *Nanuqsaurus* may have been the adaptive response by this top predator to the resource limits imposed by a highly seasonal environment (Fiorillo and Tykoski, 2014). This response would be similar to that observed with modern carnivores in an insular setting where resources are also limited compared to resource availability for mainland populations (Lomolino, 2005).

## THESCELOSAURID

"Hypsilophodontids" were a group of small, bipedal, herbivorous, basal ornithopod dinosaurs known from Middle Jurassic through Upper Cretaceous rocks in both the northern and southern hemispheres. Sternberg first distinguished morphological differences among the group, and particularly between *Thescelosaurus* and *Hypsilophodon*, and erected the family name Thescelosauridae (Sternberg, 1937). He then reconsidered and relegated the group to a lower-order taxonomic grouping, the Thescelosaurinae (Sternberg, 1940). While the "hypsilophodontids" were once considered to be a monophyletic group, more detailed work has shown the group to be paraphyletic (Scheetz, 1998, 1999; Winkler et al., 1998; Butler et al., 2008; Boyd et al., 2009), resulting in the term hypsilophodontid being abandoned as a taxonomic grouping. The family name Thescelosauridae of Sternberg (1937) was resurrected and redefined under modern ancestry-based methods to encompass a monophyletic group of small, bipedal, herbivorous Late Cretaceous North American and Asian species that had once been lumped in as "hypsilophodontids" (Brown et al., 2013).

Based on a single cheek tooth, Clemens and Nelms (1993) identified the thescelosaurid *Thescelosaurus* in the dinosaur fauna from the microvertebrate fossil site referred to as *Pediomys* Point in the Prince Creek Formation. Additional early accounts referenced multiple small ornithischian teeth as well as a toe bone attributable to the genus (Clemens, 1994; Gangloff, 1994). An additional tooth was recovered from a second site, referred to by Brown and Druckenmiller (2011) as Norton's Bed ("Norton Bed" of Fiorillo and Gangloff, 2000; Fiorillo et al., 2010a). The premaxillary teeth show great similarity to premaxillary teeth of *Thescelosaurus* (Galton, 1973;

Morris, 1976; Boyd et al., 2009), but they are also like other thescelosaurids such as *Orodromeus* (Brown and Druckenmiller, 2011). Brown and Druckenmiller (2011) compared the cheek teeth of these Alaskan thescelosaurids to both *Thescelosaurus* and *Parksosaurus*, anatomically similar taxa (Boyd et al., 2009). Their examination suggests that the Alaskan teeth are more like *Parksosaurus* than *Thescelosaurus*, but there are subtle differences.

Stratigraphic work along the Colville River on the Prince Creek Formation exposures that include where these teeth are from has shown the dinosaur-bearing rocks in the area to be early Maastrichtian in age, or approximately 70 million years old (Flores et al., 2007; Fiorillo et al., 2010a,b; Flaig et al., 2011, 2013, 2014). In their review of the "Hypsilophodontidae," which included these two dinosaurs, Weishampel and Heinrich (1992) point out that *Thescelosaurus* is known from late Maastrichtian rocks, while *Parksosaurus* is known from early Maastrichtian rock units. So, though the teeth of these taxa are not providing definitive evidence for taxonomic determination, the dinosaur they represent is almost certainly a member of the Thescelosauridae (sensu Brown et al., 2013); it may be that the stratigraphic level, and hence age, is a separate line of evidence to suggest that these teeth are more appropriately referred to as *Parksosaurus*.

## THE HADROSAUR—*EDMONTOSAURUS* OR *UGRUNAALUK*

The first dinosaurian bones found from the North Slope were those of duck-billed dinosaurs, Hadrosauridae (Davies, 1987). It seems appropriate that the first Alaskan dinosaur bones are attributable to this family given that hadrosaurs were among the first dinosaurs recognized in North America (Leidy, 1856). These animals were the most common large herbivore in the Late Cretaceous of Asia, Europe, and North America. The hadrosaurs can be subdivided into two major groups, the Lambeosaurinae, or crested hadrosaurs, and the Hadrosaurinae or Saurolophinae, or flat-headed hadrosaurs. At this time, only members of the latter group have been described from Alaska (Gangloff and Fiorillo, 2010; Mori et al., 2016; Figure 4.10).

Definitive hadrosaur bones in Alaska all come from a few remarkable bonebeds along the Colville River (Fiorillo et al., 2010a, Gangloff and Fiorillo, 2010; Flaig et al., 2014). The original discovery, a surface collection of fragmentary limb bones, ribs, and vertebrae, was made by Robert Liscomb, as discussed previously (Chapter 2). This collection of material was an important discovery because at the time it was the farthest-north documented occurrence of this group of dinosaurs (Davies, 1987). Even in the earliest phase of study, what is compelling about the material is that it indicated the presence of a bonebed with a minimum number of at least seven individuals of hadrosaur (Davies, 1987), a number that would greatly increase through years of additional work (Gangloff and Fiorillo, 2010).

With well over a dozen years of detailed excavation and mapping of the most prominent of the bonebeds along the Colville River, the Liscomb Bonebed has produced a fauna that includes several theropods (Fiorillo and Gangloff, 2000), but the bonebed is dominated by the remains of late-stage juvenile saurolophine hadrosaurs (Gangloff and Fiorillo, 2010; Mori et al., 2016). There are now thousands

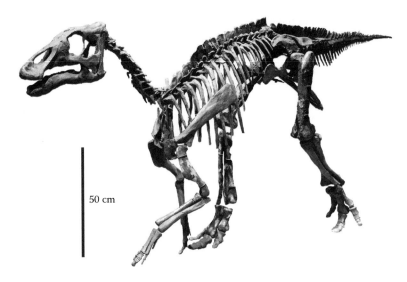

50 cm

**FIGURE 4.10** Alaskan *Edmontosaurus* sp. skeletal mount of a juvenile individual that is on display at the Perot Museum of Nature and Science. This animal is an example of a flat-headed hadrosaur.

of specimens from this site, and while there seems to be some debate regarding the fine-scale taxonomic identity of the hadrosaur (Gangloff and Fiorillo, 2010; Mori et al., 2016), there is agreement that the bonebed has produced valuable insights into the paleobiology of this ancient Arctic hadrosaur, as discussed below.

Kyle Davies' (1987) initial work attributed these bones to the Hadrosauridae, and he recognized fully grown individuals as well as juveniles, the latter determination based on the lack of fusion of the neural arches to the centra in some specimens. At the time of Davies' study, none of the diagnostic skull elements had been recovered that would have allowed a finer-scale taxonomic determination. The next phase of study by the team from the University of California–Berkeley did recover some skull elements, and they then suggested that these remains of juvenile dinosaurs could be attributed to the saurolophine, *Edmontosaurus* (Nelms, 1989). Significantly more specimens have now been collected that have provided a more complete but still imperfect understanding of this animal. Additional work suggested that attributing these bones to this genus was warranted based on aspects of the skull elements (Gangloff and Fiorillo, 2010), an assessment continued by others (Xing et al., 2014). At least with respect to the work by Gangloff and Fiorillo (2010), it was acknowledged that taxonomic identification is hampered because the best preserved taxonomically useful hadrosaurian remains all belonged to late-stage juveniles rather than adults. Taking a conservative approach, it was felt that attributing the remains to the known genus *Edmontosaurus* was appropriate, while recognizing that due to the immature growth stage of the bones a definitive identification would have to wait (Gangloff and Fiorillo, 2010). These authors also acknowledged that with the recovery of fully grown and taxonomically diagnostic individuals, this generic attribution may not stand up.

To better understand the need for an improved dataset that includes fully mature individuals of this dinosaur, consider the physical changes that familiar modern animals (including humans) go through as they age. Juvenile animals have features of their eyes and heads that are large in proportion to the rest of their bodies when compared to adult forms of the same species. These body proportions change with ontogeny from immature individuals to mature forms. Juvenile hadrosaurs share similar cranial features but because they are too young to have obtained adult features the characters defining species would not be present, appearing as the animals reached maturity.

While acknowledging that traditional methods are inconclusive, Mori and others (2016) proposed a new, different method using bivariate plots to account for ontogenetic variation observed in the Ocean Point hadrosaur that proposed to account for morphological changes as individuals reach maturity. Further, they identified a suite of characters within the skull of these animals that they argued separated this hadrosaur from *Edmontosaurus*. As a result of their analyses, Mori and others (2016) have put forth a new genus and species name *Ugrunaaluk kuukpikensis*. Interestingly, even after putting forth this new method for identifying taxa and establishing a new taxonomic name for the Arctic hadrosaur, these authors too acknowledge that adult material is required to clarify the differences between this new taxon and the known genus *Edmontosaurus*.

This point was emphasized in the study of the genus *Edmontosaurus* by Xing and others (2017). In their study, they conclude that there are two species of *Edmontosaurus*—*E. annectens* and *E. regalis*—with nine subtle but distinct morphological features of the skull separating the two species. They argued that the Alaskan material represents very young individuals that were much smaller than most comparable material available for *Edmontosaurus*. The overlap of bone proportions from known juveniles *Edmontosaurus* with those of the Alaskan material is extremely limited (in the case of *E. annectens*) or nonexistent (in the case of *E. regalis*). Xing and others (2017) therefore point out that Mori and others (2016) compared juvenile material from Alaska with adult material found elsewhere in western North America, and thus consider *Ugrunaaluk kuupikensis* invalid. In their rebuttal of Ugrunaaluk, Xing and others (2017) are suggesting that, in essence, one cannot take a photograph of a four-year-old child, skip the next twenty years of growth, take a picture of the child as a twenty-four-year old, and call that individual something new. So it seems that whether one takes a taxonomically conservative approach or attempts a more aggressive approach by applying new methods to erect new taxa, there is agreement that the best methods still involve the use of bones of fully mature individuals.

Because of the now abundant fossil record of these animals, the ancient high-latitude ecosystem recorded within the Prince Creek Formation shows the environment to have been extraordinarily rich in terms of biological productivity. The paleobiological insights from the Liscomb bonebed as well as lesser-studied neighboring bonebeds indicate the landscape was heavily populated by late-stage gregarious juvenile hadrosaurs (Fiorillo et al., 2010a; Gangloff and Fiorillo, 2010; Flaig et al., 2014).

The internal microscopic structure of bone records bone growth history. Thus, studying the histology of bones provides valuable insights into the growth patterns of vertebrates, and Chinsamy and others (2012) examined several bones of these Alaskan hadrosaurs. Their histological analysis consistently showed textural changes during growth for Alaskan hadrosaurs (Chinsamy et al., 2012). These changes probably reflect shifts in energy balance—in other words, these animals had seasonally variable growth rates through the ancient Arctic year (Chinsamy et al., 2012). These changes in growth rates presumably were an adaptive response to the pronounced annual periodic fluctuations of the light regime within the high latitudes (Chinsamy et al., 2012). Further, the bones sampled showed that the animals died coming out of a period of slow growth that presumably corresponded to the long Arctic winter (Chinsamy et al., 2012).

Evidence such as biomechanical analysis of the limb proportions of these hadrosaurs suggests that the juvenile individuals recovered from the Prince Creek Formation were likely too small to migrate ahead of the descending winter darkness (Fiorillo and Gangloff, 2001; Bell and Snively, 2008). Therefore it seems that these animals were year-round residents of the Cretaceous Arctic, so adaptations for this existence likely followed some combination of the patterns observed in modern Arctic residents who adapt either by structural adaptations (i.e., forming seasonal layers of insulation), behavioral adaptations (i.e., seeking shelter), or metabolic adaptations. For these hadrosaurs, some previous workers have suggested the hadrosaurs survived through an Arctic winter by consuming secondary food resources such as rooting for rhizomes or different vegetation entirely (Gangloff and Fiorillo, 2010; Fiorillo, 2011), and/or modifying metabolic rates so that nutritional needs were reduced during this time of the Arctic year (Gangloff and Fiorillo, 2010; Chinsamy et al., 2012). Given that the Prince Creek Formation has now been heavily sampled and studied and no large-scale burrows have yet been found (Flaig et al., 2011, 2013), it seems unlikely that these dinosaurs hibernated in burrows. Rather, it may be that the changes in metabolic rates through the year for these Arctic hadrosaurs more closely resembled the state of torpor—that state of reduced physiological activity associated with reduced food intake, commonly seen in birds (Fiorillo, 2011) and such metabolic changes occurred above ground rather than below.

## *EDMONTONIA*

The Ankylosauria comprise a group of heavily armored dinosaurs that have been affectionately nicknamed "tank dinosaurs." The Ankylosauria are often placed with another armored group, the stegosaurs, within the broader taxonomic group the Thyreophora. Within the Ankylosauria are two families: the Ankylosauridae and the Nodosauridae. The most apparent difference between them is that members of the former have a prominent tail club while members of the latter do not. Both families were herbivorous, likely slow-moving animals, and in general relatively uncommon components of the Late Cretaceous ecosystem of western North America.

One of the more historically interesting Alaskan dinosaur bone-based discoveries outside of the North Slope was the recovery of a partial skull of the nodosaurid ankylosaur *Edmontonia* (Gangloff, 1995). The specimen was found much farther to the south in Alaska in the east-central Talkeetna Mountains, which lie just north of Anchorage. This partial skull not only was the first definitive dinosaur bone found outside of northern Alaska, it also represented the first record of Ankylosauria north of Alberta, Canada. The skull, which was encased in a calcareous concretion containing fragments of the marine bivalve mollusk *Inoceramus*, eroded out of the Upper Cretaceous Matanuska Formation, a rock unit correlative with the Prince Creek Formation (Gangloff, 1995). Given the recovery of this specimen within a marine rock unit as well as several other reports of nodosaurids from marine rocks, it has been suggested these animals frequented coastal habitats, died, and were washed out to sea (Horner, 1979; Coombs and Maryanska, 1990; Gangloff, 1995).

## ALASKACEPHALE

In 1999, heavy rains on the North Slope forced a temporary closure of excavation at the Liscomb Bonebed. Long-time Arctic naturalist Dr. David Norton and I took a boat and headed up the river from the bonebed to wait out the rain. Beaching the boat on a bank of the Colville River to investigate an area known for fossil clams (Suarez et al., 2016), Norton, a research affiliate of the University of Alaska, found a bone fragment among the eroded talus at the foot of the cliffs on the riverbank. That specimen was a nearly complete left squamosal bone, from the back part of the skull, with an attached basal portion of the parietal dome of the top part of the skull, representing the first record of pachycephalosaurs in the ancient Arctic (Gangloff et al., 2005). Pachycephalosaurs are a group of bipedal, herbivorous dinosaurs with thickened skulls that are often ornamented. The ornamental nodes on the specimen Norton found have prominent apices and well-defined polygonal bases, giving the thick layer of bone a decorated appearance.

The original authors preferred a more conservative route of description because of the fragmentary nature of the specimen, and only referred the specimen to a genus that resembled *Pachycephalosaurus* (Gangloff et al., 2005). Robert Sullivan (2006), however, in reading the original paper, took a more aggressive approach. He thought that the ornamentation on the specimen, two divergent rows of nodes converging toward the midline of the skull, justified erecting a new genus and species of pachycephalosaur, *Alaskacephale gangloffi*.

## PACHYRHINOSAURUS

Horned-faced dinosaurs, or ceratopsians, are a group of dinosaurs that have been known for almost 150 years. The dramatic features of the ceratopsian skulls have captured the public's imagination, with artists often depicting the most famous of the group, *Triceratops*, pitched in a life-or-death struggle with the mighty *Tyrannosaurus*. Like the pointy-toothed predator, ceratopsians also made their screen debut in the silent film, *The Ghost of Slumber Mountain*. Not surprisingly, in the film these two dinosaurs fight to the death.

The more technical name for the group is Ceratopsia, coined by Othniel Marsh in 1890. The first recognized horned dinosaur, *Monoclonius crassus*, was discovered and named in 1876 by Edward Drinker Cope from bones collected in the Upper Cretaceous Judith River Formation of central Montana. Since the first discoveries, this fascinating group of dinosaurs has provided one of the best fossil records of any dinosaurs, though that record is largely restricted to the middle latitudes of western North America and eastern Asia, although a possible record in Australia presages their discovery more broadly across ancient geography. In general, among the more derived horned dinosaurs, the long-frilled and short-frilled varieties demonstrate their diversity, though it is the combination of features of the face and the frills that provide the most information in the study of the taxonomy of this group.

In 1988, a joint University of California–University of Alaska field party worked the Prince Creek Formation bluffs along the Colville River. While most members of the field party worked the area around the Liscomb Bonebed, Dr. Howard Hutchison of the University of California-Berkeley worked a few kilometers up river and discovered the first skull of the horned dinosaur *Pachyrhinosaurus* known from Alaska (Clemens, 1994). *Pachyrhinosaurus* belongs to the short-frilled variety of horned dinosaurs, more technically known as centrosaurines. The genus was first described based on three partial skulls found in southern Alberta by Charles M. Sternberg (1950), and named for the bizarre rugosity on the noses of these animals. Rather than the more typical facial horns, this species has enlarged nasal and supraorbital bosses—enlarged bone thickenings—on the skull. These features are the basis for the animal's name, as *Pachyrhinosaurus* means "thick-nosed lizard." The Berkeley specimen from Alaska, lacking its frill, was an isolated specimen, not part of a larger bonebed.

The best-preserved ceratopsian skeletal materials thus far recovered from the North Slope are from the Kikak-Tegoseak Quarry (Fiorillo et al., 2010b), the same quarry that produced the skull pieces of *Nanuqsaurus* (Fiorillo and Tykoski, 2014). In 1994, pieces of bone were found along the banks of the Colville River at the base of a bluff that rises over 100 meters above the river (Fiorillo et al., 2010b; Figure 4.11). In 1997, Ron Mancil, a University of Alaska student, and David Norton found the source of these bone fragments eroding out near the top of the bluff. The Kikak part of the site name is derived from the bonebed being near the confluence of the Kikak and Colville rivers, while the Tegoseak part of the name is to honor Ron Mancil's maternal grandmother.

A highly productive excavation of this site by the Perot Museum of Nature and Science from 2005 to 2007 resulted in several tons of material removed for study. The contained fossils, which include cranial elements, are from a bonebed and have been attributed to the genus *Pachyrhinosaurus* based on the thick and flattened top of the skull between orbits and the nasal openings (Fiorillo and Tykoski, 2012, 2013; Tykoski and Fiorillo, 2013; Figure 4.12). A close relative, *Achelousaurus*, is a related short-frilled ceratopsian, also with a characteristic nasal boss (Sampson, 1995), but the Kikak-Tegoseak specimens have much larger nasal bosses that extend onto the frontal bones, confirming the diagnosis for *Pachyrhinosaurus*.

The large amount of material recovered during this excavation included frill material that proved to be especially important because it provided a fuller perspective on the anatomy of this Alaskan ceratopsian. With this more complete understanding of the anatomy, a combination of features characters of this genus not seen before provided evidence for erecting the new species *Pachyrhinosaurus perotorum* (Fiorillo and Tykoski, 2012).

**FIGURE 4.11**  Excavation of Kikak-Tegoseak Quarry, Prince Creek Formation, North Slope, Alaska, with the Colville River in the background. This photograph is from the 2006 excavation, and the helicopter is preparing to sling a load containing one of the plaster field jackets containing the bones of *Pachyrhinosaurus perotorum.*

**FIGURE 4.12**  Reconstruction of *Pachyrhinosaurus perotorum.* (Artwork by Karen Carr. Full mural on display at the Perot Museum of Nature and Science, Dallas, Texas.)

A distinctive element within any ceratopsian skeleton is the occipital condyle, the bone at the base of the skull that articulates with the vertebral column. The bone resembles the ball of a trailer hitch (Figure 4.13). The side-to-side and up-and-down movement of the massive skull of a horned dinosaur is the result of that trailer hitch articulation surface. Based on the number of *Pachyrhinosaurus perotorum* occipital condyles from the Kikak-Tegoseak Quarry, the minimum number of individuals is ten (Tykoski and Fiorillo, 2013). However, the discovery of a partial juvenile skull that is lacking an occipital condyle is clearly a different-sized animal and raises the number of individuals officially to eleven (Fiorillo and Tykoski, 2013). The more immature specimen also indicates that this species lived in family groups (Fiorillo et al., 2010b).

How an animal grows is part of the basis for understanding the biology of the animal. In a preliminary study, histological evidence from these Alaskan ceratopsians indicates conspicuous growth bands, suggesting that these animals had seasonal growth, and further, that the young had rapid growth rates as young individuals (Erickson and Druckenmiller, 2011). Also with regard to these Alaskan ceratopsians, the quarry produced a young specimen that provided some interesting insights into how the remarkable nose feature of *Pachyrhinosaurus* grew through life. Comparison between this relatively juvenile specimen and the more complete skull of a mature *Pachyrhinosaurus perotorum* showed that the juvenile was

2 cm

**FIGURE 4.13** Occipital condyle of a *Pachyrhinosaurus perotorum* from the Kikak-Tegoseak Quarry. This bone is from the base of the skull and articulates with the neck. (Perot Museum of Nature and Science specimen DMNH 22257.)

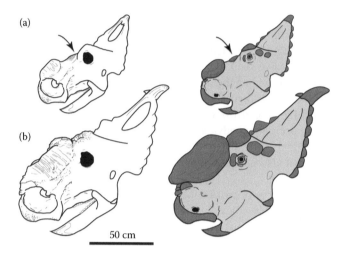

50 cm

**FIGURE 4.14**  Illustration of pattern of growth of the boss on the face of *Pachyrhinosaurus perotorum*. Light tan indicates "normal" skin, while dark brown indicates cornified scales, horns, sheaths, or other cornified tissue. The arrows indicate the dorsal rugose patch on the nasal posterior of the incipient nasal boss. This patch shows that the development of a thick cornified pad existed separate from the nasal horn sheath prior to the development of a full nasal boss at sexual maturity. See Fiorillo and Tykoski (2013) for a full discussion of the hypothesized pattern of growth. (a) Illustration of the skull of a sexually immature individual. (b) Illustration of the skull of a fully mature individual.

approximately two-thirds adult size (Fiorillo and Tykoski, 2013). Growth of the huge boss on the nose began toward the front and expanded with age toward the eyes (Fiorillo and Tykoski, 2013). *Pachyrhinosaurus* had a cornified pad on its thickened nose—something that would have appeared similar to the pad seen on the horn base of a musk ox or an African buffalo. Previous workers speculated that the cornified pad grew in *Pachyrhinosaurus* after the bony boss grew, but the Kikak-Tegoseak specimen suggests otherwise. On this Alaskan specimen, the bone surface texture on a section of the snout between the immature horn and the orbits is similar to the horn of the musk ox and buffalo, suggesting that the cornified pad was in place across the nose before the bony boss grew to its full adult size (Figure 4.14).

Because *Pachyrhinosaurus* may have used this nasal boss in shoving or head-butting matches (Sternberg, 1950) presumably during mating season, the incomplete development of the nasal boss exhibited in this juvenile specimen from the Kikak-Tegoseak Quarry would indicate that this immature individual was likely not yet sexually mature (Fiorillo and Tykoski, 2013). Thus, the recovery of a juvenile from this bonebed demonstrates this Arctic herd contained more than one age group of animal. To go one step further, a mixed age group within a herd implies some degree of care for the younger members of the Arctic *Pachyrhinosaurus* herd.

Other useful morphological information in many vertebrates lies within the brain-case and endocranial cavity, and with respect to ceratopsians, study of this region of the skull is emerging and providing additional insights into the biology of these animals. The braincase in mature ceratopsids is a robust structure with a high-fossilization

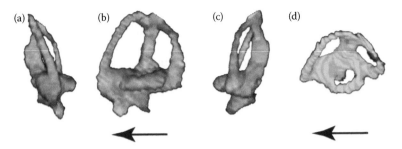

**FIGURE 4.15** Digitally rendered images of the inner ear anatomy, and specifically the semicircular canal of *Pachyrhinosaurus perotorum*. See Tykoski and Fiorillo (2013) for a full discussion. These images are for the left inner ear and provide an (a) anterior view, (b) lateral view, (c) posterior view, and (d) dorsal view. For (b) and (d), the arrows are pointing anteriorly.

potential. The first detailed description of a short-frilled, or centrosaurine, ceratopsid braincase was made by Langston (1975) in his seminal works on *Pachyrhinosaurus canadensis* (Langston, 1967, 1968, 1975). Thanks as well to the description of the braincase of the other species of this genus, *Pachyrhinosaurus lakustai* (Currie et al., 2008; Witmer and Ridgely, 2008) and *Pachyrhinosaurus perotorum*, it is clear that the braincase morphology is very similar among the three species, a not-unexpected conclusion given the close relationship between the taxa (Tykoski and Fiorillo, 2013). There are, however, at least two details in *Pachyrhinosaurus perotorum* that seem to distinguish this Alaskan species from the two Canadian species.

The first involves a deep groove on side of the braincase, referred to as the lateral surface of the basal tuber. The groove marks the path of an artery leading to the brain near the opening for a facial nerve. The path of this artery as it runs along the outside of the braincase is slightly different in the three species (Tykoski and Fiorillo, 2013).

The second feature, discernible from CT imagery, or more fully termed "computerized tomography," is within the inner ear—specifically, the semicircular canals, anatomical features that are important in controlling balance and motion in all vertebrates (Figure 4.15). The size of the ear canals among the *Pachyrhinosaurus* species is similar, yet the placement of part of one of the canals varies (Tykoski and Fiorillo, 2013). If these variations are meaningful, they probably relate to slightly different habitual head postures.

## NON-DINOSAURIAN CRETACEOUS VERTEBRATES

Another notable additional Cretaceous fossil vertebrate from northern Alaska is a single fossil turtle specimen, though this specimen is attributed to the older Nanushuk Formation rather than the younger Prince Creek Formation. This specimen is an internal mold of a shell of a turtle attributed to the family Dermatemydidae (Parrish et al., 1987). Dermatemydids are largely extinct and have only one living member, *Dermatemys mawii*, found in Central America. The fossil record of this group extends over Asia, Europe, and North America. This specimen remains the only known turtle from the Cretaceous of Alaska.

From the Prince Creek Formation, there is also fossil amphibian material. These fragments are suggestive of either a frog or a wide-mouthed salamander, with further study to determine the exact taxonomy. While the taxonomy will be important for understanding the biodiversity, either determination does not contribute to an understanding of ancient climate, as there are examples of amphibians such as the Wood Frog (*Rana sylvatica*) living well above the Arctic Circle today.

## MAMMALS

As large as the state of Alaska is, it is remarkable that the record for Mesozoic mammals in this region is known from only one site along the Colville River, *Pediomys* Point. Mammalian remains are almost exclusively isolated teeth and jaw fragments recovered by sieving. The mammalian fauna is dominated by the remains of Cretaceous marsupials (Figure 4.16) although the mammal fauna contains at least three species (Clemens and Nelms, 1993), indicating both biodiversity and unexpected taxa, along with an ecological partitioning within this group. All of these mammals had small body sizes, and because small terrestrial vertebrates do not undertake extensive annual migrations, they, like the dinosaurs, were year-round residents of the ancient North (Clemens and Nelms, 1993).

### *CIMOLODON*

*Cimolodon* is a member of the Multituberculata, an extinct order of mammals that resembled rodents, which had not yet evolved at the time the Alaska fossils lived, in form and probably habits. With a geologic history lasting approximately 120 million years, multituberculates were one of the most evolutionarily successful mammalian clades of all time. The group was widespread throughout the northern hemisphere in the Late Cretaceous, and in many localities multituberculates typically make up the majority of a mammal fauna during this window of geologic time.

The genus *Cimolodon* was named by Othniel Marsh (1889) and is known from much of the Late Cretaceous of North America. From *Pediomys* Point, partial tooth specimens resemble *Cimolodon nitidus*, an animal originally described from the

**FIGURE 4.16**   Reconstruction of a Cretaceous marsupial representative of one type of fossil mammal known from specimens from northern Alaska. (Artwork by Karen Carr.)

Cretaceous of eastern Wyoming. A second multituberculate species seems to be present but has not yet been defined (Clemens and Nelms, 1993).

## PEDIOMYS

By far the most common mammal from this locality is the taxon that is the namesake for the site, *Pediomys*, represented by more than 50 isolated molars and premolars, a maxillary fragment, and partial lower jaws, or dentaries. *Pediomys* was also named by Othniel Marsh (1889). He erected the genus based on an isolated tooth from the Maastrichtian Lance Formation of Wyoming. Several species of *Pediomys* have subsequently been named. This genus and closely related forms have risen in taxonomic status from subfamily to family level within Linnaean classification, though Davis (2007) redefined the term Pediomyidae within cladistic classification based on numerous derived characters within the teeth. Those characters reflect the shearing and crushing functions of the teeth (Davis, 2007). Pediomyids can be found in many of the latest Cretaceous mammalian faunas throughout western North America. In his review of the "pediomyids," Davis showed with one of the teeth from *Pediomys* Point that the Alaskan species is close to *Pediomys elegans* (Davis, 2007).

However, preliminary examination of the teeth also suggests that the Alaskan forms are about a quarter of the size of the southern forms. Further, these Alaskan animals likely lived in the ancient Arctic year-round. So also, considering the geographic distance between this site and the sites that produced *Pediomys elegans* in the more southern latitudes, it seems that the Alaskan pediomyid could prove to be a new taxon within the family.

The frequency of *Pediomys* teeth within the mammalian assemblage of *Pediomys* Point is particularly intriguing. In contrast, many mammalian faunas found in correlative rocks in lower latitudes are dominated by multituberculates. Is this frequency a reflection of some adaptative advantage for these marsupials compared to the more typically abundant and clearly evolutionarily successful multituberculates?

One key may be within the reproductive system of marsupials. Modern marsupial females have a double reproductive tract: two vaginas and two uteri, with the vaginas joining below the two uteri (Berra, 1998). A pseudovaginal canal forms at birth, which then allows the fetus to pass to the cloaca, and in most marsupials this pseudovaginal canal must reform with each pregnancy (Berra, 1998). The gestation period for marsupials is extremely short, perhaps as result of birth needing to take place soon after the egg membrane ruptures to avoid immunological attack (Berra, 1998). This reproduction strategy can result in greater rate of neonatal death, but with such a small energy investment given the short gestation, females sometimes breed multiple times within the same season. Strahan (1995) argues against the perceived notion of inferiority of marsupials based on their reproductive systems and suggests instead that this is a beneficial strategy. Given the long geologic record of marsupials, this argument seems to have merit. Strahan (1995) further points out modern marsupials have a lower body temperature and metabolic rate than comparably sized placental mammals—physiological features that would also seem to be advantageous when food is scarce.

## GYPSONICTOPS

Eutherians, a group of mammals that first appears some 160 million years ago in the Jurassic, comprise the group that includes mammals with placentas. At *Pediomys* Point, eutherians are represented by virtue of isolated teeth referred to the genus *Gypsonictops*, a genus named by the preeminent paleontologist and evolutionary biologist George Gaylord Simpson in 1927. While this animal first appears in the Late Cretaceous of western North America, the generic lineage continues into the Paleogene.

## FISHES

Teeth and fragmentary specimens of cartilaginous fishes, sharks and rays (chondrichthyans), and bony fishes (osteichthyans) are known from marine or near-shore deposits (Figure 4.17; Clemens and Nelms, 1993; Fiorillo and Gangloff, 2000; Fiorillo, 2006; Hilton and Grande, 2006). Only one nearly complete articulated skeleton of a bony fish is known, *Chandlerichthys strickeri*, a freshwater deep-bodied bony fish with an exceptionally small head (Grande, 1986). *Chandlerichthys* was recovered by United States Geological Survey geologist Gary Stricker from what was then known as the Chandler Formation, a rock unit name that has since been abandoned and revised to be included within the Nanushuk Formation, which spans from the Albian to the Cenomanian, or approximately 115 to 94 million years ago. (Mull et al., 2003). The specimen was recovered from a fluvial sandstone bracketed stratigraphically by coal (Grande, 1986).

Grande (1986) showed that *Chandlerichthys* belonged to the Osteoglossomorpha, which are considered among the most primitive bony fishes still living. Members of

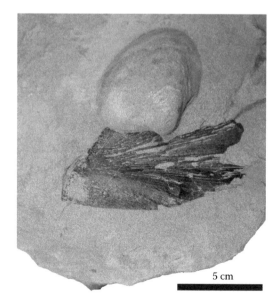

**FIGURE 4.17** Cretaceous fish skull element (subopercle) from the Schrader Bluff Formation of northern Alaska. (Perot Museum of Nature and Science specimen DMNH 18279.)

this group of fishes are notable for having toothed or bony tongues, and within the living forms the gastrointestinal tract passes to the left of the stomach, while in other fishes it passes to the right. They are all predatory fishes, and some forms use weak electrical fields to sense their prey.

Modern forms of osteoglossomorphs are mostly freshwater fishes and are found primarily in Australia, Africa, Asia, and South America, with only the genus *Hiodon* found in North America. The modern fishes range in length from approximately 2 centimeters (*Pollimyrus castelnaui*) to the largest reaching lengths as much as 2.5 meters (*Arapaima gigas*). Among the more bizarre-looking forms are the freshwater elephant fishes of the family Mormyridae, which are today found in Africa.

The oldest fossil record for osteoglossomorphs is early in the Cretaceous, approximately 125 million years ago, with a suggestion that the record may extend back into the Jurassic Period (Wilson and Murray, 2008). Molecular data from modern osteoglossomorphs suggests that the earliest osteoglossomorphs likely diverged from other more primitive fishes in the Triassic, approximately 230 million years ago (Broughton et al., 2013). In understanding fish evolutionary history, the Cretaceous is a critical time, as it is a time of transition from more primitive bony fishes to the later Paleogene bony fishes that are similar to modern bony fishes (Grande, 1986).

So, though one specimen, *Chandlerichthys* informs us that these more primitive fishes lived across a wide latitudinal range early in the known fossil history of osteoglossomorph. Further, these fishes were presumably well adapted to the seasonally extreme high-latitude environment of the ancient Arctic. With additional specimens from the Cretaceous of Alaska, we will also be better informed about the evolutionary history that led to this successful ecological exploitation.

## SUMMARY

After decades of collecting, there continues to be a growing and exciting record of bones of fossil vertebrates from Alaska that in some ways challenges what we think we know about dinosaurs. Far from complete, these specimens are providing new insights not only into biodiversity within the Arctic of the Cretaceous, but also how broad patterns of biodiversity changes carry through to deep geologic time. For example, there is a biodiversity gradient that decreases, moving from the lower latitudes to the higher latitudes.

Additionally, this growing fossil record shows these animals, rather than struggling through life in the highly seasonal polar world, were well adapted and thriving in the ancient North. Some animals, such as *Troodon*, by virtue of their large eyes, took advantage of the extreme light regime of the high latitudes and became dominant predators compared to their presence in lower latitude ecosystems, and as dominant predators they accessed food resources more readily, resulting in an increased body size compared to southern forms.

In contrast to this adaptation strategy, the tyrannosaur *Nanuqsaurus* evolved a different approach. Tyrannosaurs seem to have been near the upper limit for body size of theropods, and food intake is a main constraining factor among such predators today. Therefore by being a diminutive tyrannosaur, *Nanuqsaurus* may have been responding to the profoundly seasonal ancient Arctic environment by reducing its body size.

Among the large plant-eating dinosaurs—specifically the ceratopsians and the hadrosaurs—the histology of their bones suggests that these northern denizens varied their metabolic needs through the Arctic year. That is, during times of restricted food availability, or when optimal food choices were not available, these dinosaurs slowed their metabolic rates and reduced their food intake needs.

Additional work on polar dinosaurs will only continue to challenge what we think we know about dinosaurs. Rather than the classic model of dinosaurs based on data from the lower latitudes, the presence of such an abundant record of high-latitude dinosaurs demonstrates how evolutionarily successful this group of animals truly was during the Cretaceous.

## REFERENCES

Abler, W.L. 1992. The serrated teeth of tyrannosaurid dinosaurs, and biting structures in other animals. *Paleobiology* 18:161–183.

Adams, T.L. 2009. Deposition and taphonomy of the Hound Island Late Triassic vertebrate fauna: Fossil preservation within subaqueous gravity flows. *Palaios* 24:603–615.

Angielczyk, K.D., RW. Burroughs, C.R. Feldman. 2015. Do turtles follow the rules? Latitudinal gradients in species richness, body size, and geographic range area of the world's turtles. *Journal of Experimental Zoology Part B: Molecular and Developmental Evolution* 324:270–294.

Ashton, K.G. 2002a. Do amphibians follow Bergmann's rule? *Canadian Journal of Zoology* 80:708–716.

Ashton, K.G. 2002b. Patterns of within-species body size variation of birds: Strong evidence for Bergmann's rule. *Global Ecology and Biogeography* 11:505–523.

Bee, J.W. and E.R. Hall. 1956. Mammals of northern Alaska on the Arctic Slope. *University of Kansas Museum of Natural History Miscellaneous Publication*, 8:1–309.

Bell, P.R. and E. Snively. 2008. Polar dinosaurs on parade: A review of dinosaur migration. *Alcheringa* 32:271–284.

Berra, T.M. 1998. *A Natural History of Australia.* Berkeley, CA: University of California Press.

Blanchard, B.M. 1987. Size and growth patterns of the Yellowstone grizzly bear. In *Bears—Their Biology and Management: Proceedings of the Seventh International Conference on Bear Research and Management*, Williamsburg, VA, ed. P. Zager, 99–107. February 21–26, 1986, and Plitvice Lakes, Yugoslavia, March 2–5, 1986. Washington, DC: International Association for Bear Research and Management.

Boyd, C., C.M. Brown, R.D. Scheetz, and J.A. Clarke. 2009. Taxonomic revision of the Basal Neornithischian Taxa Thescelosaurus and Bugenasaura. *Journal of Vertebrate Paleontology* 29:758–770.

Broughton, R.E., R. Betancur-R, C. Li, G. Arratia, and G. Ortí. 2013. Multi-locus phylogenetic analysis reveals the pattern and tempo of bony fish evolution. *PLoS Currents* 5 April 16. Edition 1. ecurrents.tol.2ca8041495ffafd0c92756e75247483e. http://doi.org/10.1371/currents.tol.2ca8041495ffafd0c92756e75247483e.

Brown, C.M. and P. Druckenmiller. 2011. Basal ornithopod (Dinosauria: Ornithischia) teeth from the Prince Creek Formation (early Maastrichtian) of Alaska. *Canadian Journal of Earth Sciences* 48:1342–1354.

Brown, C.M. D.C. Evans, M.J. Ryan, and A.P. Russell. 2013. New data on the diversity and abundance of small-bodied ornithopods (Dinosauria, Ornithischia) from the Belly River Group (Campanian) of Alberta. *Journal of Vertebrate Paleontology* 33:495–520.

Brusatte, S.L. 2012. *Dinosaur Paleobiology.* Hoboken, NJ: John Wiley & Sons, 322p.

Brusatte, S.L., M.A. Norell, T.D. Carr, G.M. Erickson, J.R. Hutchinson, A.M. Balanoff, G.S. Bever, J.N. Choiniere, P.J. Makovicky, and X. Xu. 2010. Tyrannosaur paleobiology: New research on ancient exemplar organisms. *Science* 329:1481–1485.

Butler, R.J., P. Upchurch, and D.B. Norman. 2008. The phylogeny of the ornithischian dinosaurs. *Journal of Systematic Palaeontology* 6:1–40.

Carbone, C., A. Teacher, and J.M. Rowcliffe, 2007. The costs of carnivory. *PLoS Biology* 5(2), e22p. doi:10.1371/journal.pbio.0050022.

Carbone, C., N. Pettorelli, and P.A. Stephens. 2011 Intra-guild competition and its implications for one of the biggest terrestrial predators, Tyrannosaurus rex. *Proceedings of the Royal Society B* 278:2682–2690.

Carpenter, K. 2013. A closer look at the hypothesis of scavenging versus predation by *Tyrannosaurus rex.* In *Tyranosaurid Paleobiology,* eds. J.M. Parrish, R.E. Molnar, P.J. Currie, and E.B. Koppelhus, 265–277. Bloomington, IN: Indiana University Press.

Chinsamy, A., D.B. Thomas, A.R. Tumarkin-Deratzian, and A.R. Fiorillo. 2012. Hadrosaurs were perennial polar residents. *The Anatomical Record* 295:610–614.

Chure, D.J. 1994. *Koparion douglassi,* a new dinosaur from the Morrison Formation (Upper Jurassic) of Dinosaur National Monument; the oldest Troodontid (Theropoda: Maniraptora). *Brigham Young University Geology Studies* 40:11–15.

Clark, T.W., A.P. Curlee, S.C. Minta, and P.M. Kareiva (eds.). 1999. *Carnivores in Ecosystems: The Yellowstone Experience.* New Haven, CT: Yale University Press.

Clarke, A. and K.J. Gaston. 2006. Climate, energy, and diversity. *Proceedings of the Royal Society B* 273:2257–2266.

Clemens, W.A. 1994. Continental vertebrates from the Late Cretaceous of the North Slope, Alaska. In *1992 Proceedings International Conference on Arctic Margins: Outer Continental Shelf Study,* eds. D.K. Thurston and K. Fujita, 395–398. Mineral Management Service 94-0040.

Clemens, W.A. and L.G. Nelms. 1993. Paleoecological implications of Alaskan terrestrial vertebrate fauna in latest Cretaceous time at high paleolatitudes. *Geology* 21:503–506.

Colbert, E. and D.A. Russell. 1969. The small Cretaceous dinosaur Dromaeosaurus. *American Museum Novitates* 2380:1–49.

Coombs, W.P. and T. Maryanska. 1990. Ankylosauria. In *The Dinosauria,* eds. D.B. Weishampel, P. Dodson, and H. Osmolska, 456–483. Berkeley, CA: University of California Press.

Corfield, J.R., J.M. Wild, M.E. Hauber, S. Parsons, and M.F. Kubke. 2008. Evolution of brain size in the Palaeognath lineage, with an emphasis on New Zealand ratites. *Brain, Behavior and Evolution* 71:87–99.

Crabtree, R.L. and J.W. Sheldon. 1999. Coyotes and canid coexistence in Yellowstone. In *Carnivores in Ecosystems: The Yellowstone Experience,* eds. T.W. Clark, A.P. Curlee, S.C. Minta, and P.M. Kareiva, 127–163. New Haven, CT: Yale University Press.

Currie, D.J. 1991. Energy and large-scale patterns of animal and plant-species richness. *American Naturalist* 137:27–49.

Currie, P.J. 1985. Cranial anatomy of *Stenonychosaurus inequalis* (Saurischia, Theropoda) and its bearing on the origin of birds. *Canadian Journal of Earth Sciences* 22:1643–1658.

Currie, P.J. 1987. Bird-like characteristics of the jaws and teeth of troodontid theropods (Dinosauria, Saurischia). *Journal of Vertebrate Paleontology* 7:72–81.

Currie, P.J. 1995. New information on the anatomy and relationships of the *Dromaeosaurus albertensis* (Dinosauria: Theropoda). *Journal of Vertebrate Paleontology* 15:576–591.

Currie, P.J., J.K. Rigby, Jr., and R. Sloan. 1990. Theropod teeth from the Judith River Formation of southern Alberta, Canada. *Dinosaur Systematics: Approaches and Perspectives,* eds. K. Carpenter and P.J. Currie, 107–125pp. New York, NY: Cambridge University Press.

Currie, P.J., W. Langston, Jr., and D.H. Tanke. 2008. *A New Horned Dinosaur from an Upper Cretaceous Bone Bed in Alberta.* Ottawa, Canada: NRC Research Press.

Davies, K.L. 1987. Duck-bill dinosaurs (Hadrosauridae, Ornithischia) from the North Slope of Alaska. *Journal of Paleontology* 61:198–200.

Davis, B.M. 2007. A revision of "pediomyid" marsupials from the Late Cretaceous of North America. *Acta Palaeontologica Polonica* 52:217–256.

DeConto, R.M., W.W. Hay, S.L. Thompson, and J. Bergengren. 1999. Late Cretaceous climate and vegetation interactions: Cold continental interior paradox. In *Evolution of the Cretaceous Ocean-Climate System*, eds. E. Barrera and C.C. Johnson, 391–406. Boulder, CO: Geological Society of America Special Paper, 332.

Druckenmiller, P.S. and E.E. Maxwell. 2014. A Middle Jurassic (Bajocian) ophthalmosaurid (Reptilia, Ichthyosauria) from the Tuxedni Formation, Alaska and the early diversification of the clade. *Geological Magazine* 151:41–48.

Druckenmiller, P.S., N. Kelley, M.T. Whalen, C. McRoberts, and J.G. Carter. 2014. An Upper Triassic (Norian) ichthyosaur (Reptilia, Ichthyopterygia) from northern Alaska and dietary insight based on gut contents. *Journal of Vertebrate Paleontology* 34:1460–1465.

Edwards, M.B., R. Edwards, and E.H. Colbert. 1978. Carnosaurian footprints in the Lower Cretaceous of Eastern Spitsbergen. *Journal of Paleontology* 52:940–941

Erickson, G.M. and P.S. Druckenmiller. 2011. Longevity and growth rate estimates for a polar dinosaur: A *Pachyrhinosaurus* (Dinosauria: Neoceratopsia) specimen from the North Slope of Alaska showing a complete developmental record. *Historical Biology* 23:327–334.

Evans, D.C., M.J. Vavrek, D.R. Braman, N.E. Campione, T.A. Dececchi, and G.D. Zazula. 2012. Vertebrate fossils (Dinosauria) from the Bonnet Plume Formation, Yukon Territory, Canada. *Canadian Journal of Earth Sciences* 49:396–411.

Evans, K.L., J.J.D. Greenwood, and K.J. Gaston. 2005. Dissecting the species-energy relationship. *Proceedings of the Royal Society B* 272:2155–2163.

Fiorillo, A.R. 1998a. Measuring fossil reworking within a fluvial system: An example from the Hell Creek Formation (Upper Cretaceous) of eastern Montana. In *Advances in Vertebrate Paleontology and Geochronology*, eds. Y. Tomida, L.J. Flynn, and L.L. Jacobs. Tokyo, Japan: *National Science Museum Monographs* 14:243–251.

Fiorillo, A.R. 1998b. Dental microwear patterns of the sauropod dinosaurs Camarasaurus and Diplodocus: Evidence for resource partitioning in the Late Jurassic of North America. *Historical Biology* 13:1–16.

Fiorillo, A.R. 2006. Review of the Dinosaur Record of Alaska with comments regarding Korean Dinosaurs as comparable high-latitude fossil faunas. *Journal of Paleontological Society of Korea* 22:15–27.

Fiorillo, A.R. 2008. On the occurrence of exceptionally large teeth of *Troodon* (Dinosauria: Saurischia) from the Late Cretaceous of northern Alaska. *Palaios* 23:322–328.

Fiorillo, A.R., 2011. Microwear patterns on the teeth of northern high latitude hadrosaurs with comments on microwear patterns in hadrosaurs as a function of latitude and seasonal ecological constraints. *Palaeontologia Electronica* 14(3):20A.

Fiorillo, A.R. and P.J. Currie. 1994. Theropod teeth from the Judith River Formation (Upper Cretaceous) of south-central Montana. *Journal of Vertebrate Paleontology* 14:74–80.

Fiorillo, A.R. and R.A. Gangloff. 2000. Theropod teeth from the Prince Creek Formation (Cretaceous) of northern Alaska, with speculations on arctic dinosaur paleoecology. *Journal of Vertebrate Paleontology* 20:675–682.

Fiorillo, A.R. and R.A. Gangloff. 2001. The caribou migration model for Arctic hadrosaurs (Ornithischia: Dinosauria): A reassessment. *Historical Biology* 15:323–334.

Fiorillo, A.R., P.J. McCarthy, and P.P. Flaig. 2010a. Taphonomic and sedimentologic interpretations of the dinosaur-bearing Upper Cretaceous strata of the Prince Creek Formation, northern Alaska: Insights from an ancient high-latitude terrestrial ecosystem. *Palaeogeography, Palaeoclimatology, Palaeoecology* 295:376–388.

Fiorillo, A.R., P.J. McCarthy, and P.P. Flaig. 2016. A multi-disciplinary perspective on habitat preferences among dinosaurs in a Cretaceous Arctic greenhouse world, North Slope, Alaska (Prince Creek Formation: Lower Maastrichtian). *Palaeogeography, Palaeoclimatology, Palaeoecology* 441:377–389.

Fiorillo, A.R., P.J. McCarthy, P.P. Flaig, E. Brandlen, D.W. Norton, P. Zippi, L.L. Jacobs, and R.A. Gangloff. 2010b. Paleontology and paleoenvironmental interpretation of the Kikak-Tegoseak Quarry (Prince Creek Formation: Late Cretaceous), northern Alaska: A multi-disciplinary study of a high-latitude ceratopsian dinosaur bonebed. In *New Perspectives on Horned Dinosaurs*, eds. M.J. Ryan, B.J. Chinnery-Allgeier, and D.A. Eberth, 456–477. Bloomington, IN: Indiana University Press.

Fiorillo, A.R., R.S. Tykoski, P.J. Currie, P.J. McCarthy, and P. Flaig. 2009. Description of two partial Troodon braincases from the Prince Creek Formation (Upper Cretaceous), North Slope Alaska. *Journal of Vertebrate Paleontology* 29:178–187.

Fiorillo, A.R. and R.S. Tykoski. 2012. A new species of centrosaurine ceratopsid Pachyrhinosaurus from the North Slope (Prince Creek Formation: Maastrichtian) of Alaska. *Acta Palaeontologica Polonica* 57:561–573.

Fiorillo, A.R. and R.S. Tykoski. 2013. An immature *Pachyrhinosaurus perotorum* (Dinosauria: Ceratopsidae) nasal reveals unexpected complexity of craniofacial ontogeny and integument of Pachyrhinosaurus. *PLoS ONE* 8(6):e65802. doi:10.1371/journal.pone.0065802.

Fiorillo A.R. and R.S. Tykoski. 2014. A diminutive new tyrannosaur from the top of the world. *PLoS ONE* 9(3):e91287. doi:10.1371/journal.pone.0091287.

Flaig, P.P., A.R. Fiorillo, and P.J. McCarthy. 2014. Dinosaur-bearing hyperconcentrated flows of Cretaceous Arctic Alaska: Recurring catastrophic event beds on a distal paleopolar coastal plain. *Palaios* 29:594–611.

Flaig, P.P., P.J. McCarthy, and A.R. Fiorillo. 2011. A tidally influenced, high-latitude coastal-plain: The upper Cretaceous (Maastrichtian) Prince Creek Formation, North Slope, Alaska. In *From River to Rock Record: The Preservation of Fluvial Sediments and Their Subsequent Interpretation*, eds. S.K. Davidson, C.P. North, and S. Leleu, 233–264. Tulsa, OK: SEPM Special Publication 97.

Flaig, P.P., P.J. McCarthy, and A.R. Fiorillo. 2013. A tidally influenced, high-latitude alluvial/coastal plain: The Late Cretaceous (Maastrichtian) Prince Creek Formation, North Slope, Alaska. In *New Frontiers in Paleopedology and Terrestrial Paleoclimatology—Paleosols and Soil Surface Analog Systems*, eds. S.G. Driese, L.C. Nordt, and P.J. McCarthy, 179–230. Tulsa, OK: SEPM Special Publication 104.

Flores, R.M., M.D. Myers, D.W. Houseknecht, G.D. Stricker, D.W. Brizzolara, T.J. Ryherd, and K.I. Takahashi. 2007. Stratigraphy and facies of Cretaceous Schrader Bluff and Prince Creek Formations in Colville River Bluffs, North Slope, Alaska. *United States Geological Survey Professional Paper* 1748:1–52.

Fortelius, M. 1990. Problems with using fossil teeth to estimate body sizes of extinct mammals. In *Body Size in Mammalian Paleobiology: Estimation and Biological Implications*, eds. J. Damuth and B.J. MacFadden, 207–228. Cambridge, UK: Cambridge University Press.

Freeman, P.W. 1981. Correspondence of food habits and morphology in insectivorous bats. *Journal of Mammalogy* 62:166–173.

Galton, P. 1973. Redescription of the skull and mandible of Parksosaurus from the Late Cretaceous with comments on the family Hypsilophodontidae (Ornithischia). *Royal Ontario Museum, Life Sciences Contribution* 89, 1–21.

Gangloff, R.A. 1994. The record of Cretaceous dinosaurs in Alaska: An overview. In *1992 Proceedings International Conference on Arctic Margins. Outer Continental Shelf Study*, eds. D.K. Thurston and K. Fujita, 399–404. Mineral Management Service 94-0040.

Gangloff, R.A. 1995. *Edmontonia* sp., the first record of an ankylosaur from Alaska. *Journal of Vertebrate Paleontology* 15:195–200.

Gangloff, R.A. 1998. Arctic dinosaurs with emphasis on the Cretaceous record of Alaska and the Eurasian-North American connection. In *Lower and Middle Cretaceous Terrestrial Ecosystems*, eds. S.G. Lucas, J.I. Kirkland, and J.W. Estep, 211–220. Albuquergue, NM: New Mexico Museum of Natural History and Science Bulletin No. 14.

Gangloff, R.A., A.R. Fiorillo, and D.W. Norton. 2005. The first Pachycephalosaurine (Dinosauria) from the Paleo-Arctic of Alaska and its paleogeographic implications. *Journal of Paleontology* 79:997–1001.

Gangloff, R.A. and A.R. Fiorillo. 2010. Taphonomy and paleoecology of a bonebed from the Prince Creek Formation, North Slope, Alaska. *Palaios* 25:299–317.

Geist, V. 1987. Bergmann's rule is invalid. *Canadian Journal of Zoology* 65:1035–1038.

Gilmore, C.W. 1924. On Troodon validus, an ornithopodous dinosaur from the Belly River Cretaceous of Alberta, Canada. *University of Alberta, Department of Geology Bulletin* 1:1–43.

Godefroit, P., L. Golovneva, S. Shchepetov, G. Garcia, and P. Alekseev. 2009. The last polar dinosaurs: High diversity of latest Cretaceous arctic dinosaurs in Russia. *Naturwissenschaften* 96:495–501.

Goswami, A., J.J. Flynn, L. Ranivoharimanana, and A.R. Wyss. 2005. Dental microwear in Triassic amniotes: Implications for paleoecology and masticatory mechanics. *Journal of Vertebrate Paleontology* 25:320–329.

Grande, L. 1986. The first articulated freshwater teleost fish from the Cretaceous of North America. *Palaeontology* 29:365–371.

Grant, P.R. 1986. *Ecology and Evolution of Darwin's Finches.* Princeton, NJ: Princeton University Press.

Hawkins, B.A., E.E. Porter, and J.A.F. Diniz-Filho. 2003. Productivity and history as predictors of the latitudinal diversity gradient in terrestrial birds. *Ecology* 84:1608–1623.

Hilderbrand, G.V., C.C. Schwartz, C.T. Robbins, M.E. Jacoby, T.A. Hanley, S.M. Arthur, and C. Servheen. 1999. The importance of meat, particularly salmon, to body size, population productivity, and conservation of North American brown bears. *Canadian Journal of Zoology* 77: 132–138.

Hilton, E.J. and L. Grande. 2006. Review of the fossil record of sturgeons, family Acipenseridae (Actinopterygii: Acipenseriformes), from North America. *Journal of Paleontology* 80:672–683.

Holtz, T.R., Jr., D.L. Brinkman, and C.L. Chandler. 1998. Denticle morphometrics and a possible omnivorous feeding habit for the theropod dinosaur Troodon. *Gaia* 15:159–166.

Holtz, T.R., Jr., R.E. Molnar, and P.J. Currie, 2004. Basal Tetanurae. In *The Dinosauria*, eds. D.B. Weishampel, P. Dodson, and H. Osmolska, 71–110. Berkeley, CA: University of California Press.

Horner, J.R. 1979. Upper Cretaceous dinosaurs from the Bearpaw Shale (marine) of south-central Montana with checklist of Upper Cretaceous dinosaur remains from marine sediments in North America. *Journal of Paleontology* 53:566–577.

Hotton, N., III, E.C. Olson, and R. Beerbower. 1997. The amniote transition and the discovery of herbivory. In *Amniote Origins: Completing the Transition to Land*, eds. S.S. Sumida and K.L.M. Martin, 207–264. San Diego, CA: Academic Press.

Hurum, J.H., J. Milàn, Ø. Hammer, I. Midtkandal, H. Amundsen, and B. Sæther. 2006. Tracking polar dinosaurs—New finds from the Lower Cretaceous of Svalbard. *Norwegian Journal of Geology* 86:397–402.

Huston, M.A. 1979. A general hypothesis of species diversity. *American Naturalist* 113:81–101.

Hutchinson, G.E. 1978. *An Introduction to Population Ecology.* New Haven, CT: Yale University Press.

Irving, L. 1972. *Arctic Life of Birds and Mammals, Including Man.* New York, NY: Springer-Verlag.

Jefferies, R.L., J. Svoboda, G. Henry, M. Raillard, and R. Ruess. 1992. Tundra grazing systems and climate change. In *Arctic Ecosystems in a Changing Climate: An Ecophysiological Perspective,* eds. F.S. Chapin, III, R.L. Jefferies, J.F. Reynolds, G.R. Shaver, and J. Svoboda, 391–412. San Diego, CA: Academic Press.

Johnson, W.E., T.K. Fuller, and W.L. Franklin. 1996. Sympatry in canids: A review and assessment. In *Carnivore Behavior, Ecology and Evolution, Vol. 2,* ed. J.L. Gittleman, 189–218. Ithaca, NY: Cornell University Press.

Kay, R.F. and H.H. Covert. 1983. True grit: A microwear experiment. *American Journal of Physical Anthropology* 61:33–38.

Kays, R.W. and D.E. Wilson. 2002. *Mammals of North America.* Princeton, NJ: Princeton University Press.

Kobayashi, Y., J.-C. Lü, Z.-M. Dong, R. Barsbold, Y. Azuma, and Y. Tomida. 1999. Herbivorous diet in an ornithomimid dinosaur. *Nature* 402:480–481.

Kobayashi, Y. and J.-C Lü. 2003. A new ornithomimid dinosaur with gregarious habits from the Late Cretaceous of China. *Acta Palaeontologica Polonica* 48:235–259.

Langston, W., Jr. 1967. The thick-headed ceratopsian dinosaur Pachyrhinosaurus (Reptilia: Ornithischia), from the Edmonton Formation near Drumheller, Canada. *Canadian Journal of Earth Sciences* 4:171–186.

Langston, W., Jr. 1968. A further note on Pachyrhinosaurus (Reptilia: Ceratopsia). *Journal of Paleontology* 42:1303–1304.

Langston, W., Jr. 1975. The ceratopsian dinosaurs and associated lower vertebrates from the St. Mary River Formation (Maestrichtian) at Scabby Butte, southern Alberta. *Canadian Journal of Earth Sciences* 12:1576–1608.

Lapparent, A.F. de. 1962. Footprints of dinosaur in the Lower Cretaceous of Vestspitsbergen-Svalbard. *Norsk Polarinstitutt Årbok* 1960:14–21.

Lawver, L.A., A. Grantz, and L.M. Gahagan. 2002. Plate kinematic evolution of the present Arctic Region since the Ordovician. In *Tectonic Evolution of the Bering Shelf-Chukchi Sea-Arctic Margin and Adjacent Landmasses,* eds. E.L. Miller, A. Grantz, and S.L. Klemperer, 333–358. Boulder, CO: Geological Society of America Special Paper 360.

Lee, Y.-N., R. Barsbold, P.J. Currie, Y. Kobayashi, H.-J. Lee, P. Godefroit, F. Escuillié, and C. Tsogtbaatar. 2014. Resolving the long-standing enigmas of a giant ornithomimosaur *Deinocheirus mirificus. Nature* 515:257–260.

Leidy, J. 1856. Notices of the remains of extinct reptiles and fishes discovered by Dr. F.V. Hayden in the badlands of the Judith River, Nebraska Territory. *Proceedings of the Academy of Natural Sciences of Philadelphia* 8:72–73.

Loewen, M.A., R.B. Irmis, J.J.W. Sertich, P.J. Currie, and S.D. Sampson. 2013. Tyrant dinosaur evolution tracks the rise and fall of Late Cretaceous oceans. *PLoS ONE* 8:1–14 (Issue 11), e79420. doi:10.1371/journal.pone.0079420.

Lomolino, M.V. 2005. Body size evolution in insular vertebrates. *Journal of Biogeography* 32:1683–1699.

Mackie, R.J. 1970. Range ecology and relations of mule deer, elk, and cattle in the Missouri River Breaks, Montana. *Wildlife Monographs* 20:1–79.

Marsh, O.C. 1889. Discovery of Cretaceous mammalia. *American Journal of Science* 38:81–92.

Mateo, J.A. and L.F. Lopez-Jurado. 1997. Dental ontogeny in *Lacerta lepida* (Sauria, Lacertidae) and its relationship to diet. *Copeia* 1997:461–463.

McFeeters, B., M.J. Ryan, C. Schröder-Adams, and T.M. Cullen. 2016. A new ornithomimid theropod from the Dinosaur Park Formation of Alberta, Canada. *Journal of Vertebrate Paleontology* 36:e1221415.

McFeeters, B., M.J. Ryan, C. Schröder-Adams, and P.J. Currie. 2017. First North American occurrences of Qiupalong (Theropoda: Ornithomimidae) and the palaeobiogeography of derived ornithomimids. *FACETS* 2:355–373.

McKane, R.B., L.C. Johnson, G.R. Shaver, K.J. Nadelhoffer, E.B. Rastetter, B. Fry, A.E. Giblin et al. 2002. Resource-based niches provide a basis for plant species diversity and dominance in arctic tundra. *Nature* 415:68–71.

McNab, B.K. 1971. On the ecological significance of Bermann's rule. *Ecology* 52:845–854.

Mittelbach, G.G., C.F. Steiner, S.M. Scheiner, K.L. Gross, H.L. Reynolds, R.B. Waide, M.R. Willig, S.I. Dodson, and L. Gough. 2001. What is the observed relationship between species richness and productivity? *Ecology* 82:2381–2396.

Mori, H., P.S. Druckenmiller, and G.M. Erickson. 2016. A new Arctic hadrosaurid from the Prince Creek Formation (lower Maastrichtian) of northern Alaska. *Acta Palaeontologica Polonica* 61:15–32.

Morris, W.J. 1976. Hypsilophodont dinosaurs: A new species and comments on their systematics. In *Athlon, Essays on Palaeontology in Honour of Loris Shano Russell*, ed. C.S. Churcher, 93–113. Toronto, ON: Royal Ontario Museum Publications in Life Sciences.

Mull, C.G., D.W. Houseknecht, and K.J. Bird. 2003. Revised Cretaceous and Tertiary Stratigraphic Nomenclature in the Colville Basin, Northern Alaska. *United States Geological Survey Professional Paper 1673*. Version 1.0 http://pubs.usgs.gov/pp/p1673/index.html.

Nelms, L.G. 1989. Late Cretaceous dinosaurs from the North Slope of Alaska. *Journal of Vertebrate Paleontology* 9 (Suppl 3):34A.

Nopsca, F. 1901. Synopsis and Abstammung der Dinosaurier. *Foldtani kolzony* 31(Suppl.): 247–288.

Osborn, H.F. 1905. Tyrannosaurus and other Cretaceous carnivorous dinosaurs. *Bulletin of the American Museum of Natural History* 21:259–265.

Parrish, J.M., J.T. Parrish, J.H. Hutchison, and R.A Spicer. 1987. Late Cretaceous vertebrate fossil from the North Slope of Alaska and implications for dinosaur ecology. *Palaios* 2:377–389.

Parrish, J.T., M.L. Droser, and D.J. Bottjer. 2001. A Triassic upwelling zone: The Shublik Formation, arctic Alaska, USA. *Journal of Sedimentary Research* 71:272–285.

Pasch, A.D. and May, K.C. 1997. First occurrence of a hadrosaur (Dinosauria) from the Matanuska Formation (Turonian) in the Talkeetna Mountains of south-central Alaska. In *Short Notes on Alaska Geology, 1997*, eds. J.G. Clough and F. Larson, 99–109. Fairbanks, AK: Alaska Department of Natural Resources, Professional Report 118.

Patton, W.W. and I.L. Tailleur. 1964. Geology of the Killik-Itkillik Region, Alaska. *United States Geological Survey Professional Paper* 303-G:409–500.

Pianka, E.R. 1966. Latitudinal gradients in species diversity: A review of concepts. *American Naturalist* 100:33–46.

Pianka, E.R., R.B. Huey, and L.R. Lawlor. 1979. Niche segregation in desert lizards In *Analysis of Ecological Systems*, eds. D.J. Horn, G.R. Stairs, and R.D. Mitchell, 67–115. Columbus, OH: Ohio State University Press.

Pigliucci, M. 2001. *Phenotypic Plasticity: Beyond Nature and Nurture*. Baltimore, MD: Johns Hopkins University Press.

Pisanty-Baruch, I., J. Barr, E.B. Wiken, and D.A. Gauthier. 1999. Reporting on North America: Continental connections. *The George Wright Forum* 16:22–36.

Rich, T.H., P. Vickers-Rich, and R.A. Gangloff. 2002. Polar dinosaurs. *Science* 295:979–980.

Rivals, F. and B. Deniaux. 2003. Dental microwear analysis for the diet of an argali population (*Ovis ammon* Antiqua) of mid-Pleistocene age, Caune de l'Arago cave, eastern Pyrenees, France. *Palaeogeography, Palaeoclimatology, Palaeoecology* 193:443–455.

Rouse, G.E. and S.K. Svrivastava. 1972. Palynological zonation of Cretaceous and Early Tertiary rocks of the Bonnet Plume Formation Northeastern Yukon, Canada. *Canadian Journal of Earth Sciences* 9:1163–1179.

Rowe, T., R.L. Cifelli, T.M. Lehman, and A. Weil. 1992. The Campanian Terlingua local fauna, with a summary of other vertebrates from the Aguja Formation, Trans-Pecos Texas. *Journal of Vertebrate Paleontology* 12:472–493.

Rosenzweig, M.L. 1968. The strategy of body size in mammalian carnivores. *American Midland Naturalist* 80:299–315.

Russell, D.A. and R. Seguin. 1982. Reconstructions of the small Cretaceous theropod *Stenonychosaurus inequalis* and a hypothetical dinosauroid. *Syllogeus* 37:1–43.

Russell, L.S. 1948. The dentary of *Troodon*, a genus of theropod dinosaur. *Journal of Paleontology* 22:625–629.

Sampson, S.D. 1995. Two new horned dinosaurs from the Upper Cretaceous Two Medicine Formation of Montana; with a phylogenetic analysis of the centrosaurinae (Ornithischia: Ceratopsidae). *Journal of Vertebrate Paleontology* 15:743–760.

Sankey, J.T. 2001. Late Campanian dinosaurs, Aguja Formation, Big Bend, Texas. *Journal of Paleontology* 75:208–215.

Sankey, J.T., B.R. Standhardt, and J.A. Schiebout. 2005. Theropod teeth from the Upper Cretaceous (Campanian-Maastrichtian), Big Bend National Park, Texas. In *The Carnivorous Dinosaurs*, ed. K. Carpenter, 127–152. Indianapolis, IN: Indiana University Press.

Schall, J.J. and E.R. Pianka. 1977. Species densities of reptiles and amphibians on the Iberian Peninsula. *Donana, Acta Vertebrata* 4:27–34.

Schall, J.J. and E.R. Pianka. 1978. Geographical trends in numbers of species. *Science* 201:679–686.

Scheetz, R.D. 1998. Phylogeny of basal ornithopod dinosaurs and the dissolution of the Hypsilophodontidae. *Journal of Vertebrate Paleontology* 18(Suppl):75A.

Scheetz, R.D. 1999. Osteology of *Orodromeus makelai* and the phylogeny of basal ornithopod dinosaurs. *PhD thesis*, Department of Geology, Montana State University, Bozeman, Montana, 186pp.

Schmitz, O.J. and D.M. Lavigne. 1987. Factors affecting body size in sympatric Ontario Canis. *Journal of Mammalogy* 68:92–99.

Scholander, P.F 1955. Evolution of climatic adaptation in homeotherms. *Evolution* 9:15–26.

Scholander, P.F. 1956. Climatic rules. *Evolution* 10:339–340.

Schluter, D. and R.E. Ricklefs. 1993. Species diversity, an introduction to the problem. In *Species Diversity in Ecological Communities*, eds. R.E. Ricklefs and D. Schluter, 1–10. Chicago, IL: University of Chicago Press.

Servheen, C. 1984. The status of the grizzly bear and the interagency grizzly bear recovery effort. In *Proceedings of the 64th Annual Conference of the Western Association of Fish and Wildlife Agencies,* July 16–19, 1984, Victoria, British Columbia, 227–234.

Spicer, R.A. 2003. Changing climate and biota. In *The Cretaceous World*, ed. P. Skelton, 85–162. Cambridge, UK: Cambridge University Press.

Sternberg, C.M. 1937. *Classification of Thescelosaurus, with a Description of a New Species* (Abstract 375). *Proceedings for the Geological Society of America 1936.*

Sternberg, C.M. 1940. *Thescelosaurus edmontonensis*, n. sp., and classification of the Hypsilophodontidae. *Journal of Paleontology* 14:481–494.

Sternberg, C.M. 1950. *Pachyrhinosaurus canadensis*, representing a new family of the Ceratopsia, from southern Alberta. *Bulletin of the National Museum of Canada* 118:109–120.

Strahan, R. 1995. *The Mammals of Australia*, 2nd edition. Chatswood, NSW: Reed Books.

Suarez, C.A., P.P. Flaig, G.A. Ludvigson, L.A. González, R. Tian, H. Zhou, P.J. McCarthy, D.A. Van der Kolk, and A.R. Fiorillo. 2016. Reconstructing the paleohydrology of a cretaceous Alaskan paleopolar coastal plain from stable isotopes of bivalves. *Palaeogeography, Palaeoclimatology, Palaeoecology* 441:339–351.

Sullivan, R.M. 2006. A taxonomic review of the Pachycephalosauridae (Dinosauria: Ornithischia). In *Late Cretaceous Vertebrates from the Western Interior*, eds. S.G. Lucas and R.M. Sullivan, 347–3365. Albuquerque, NM: New Mexico Museum of Natural History and Science Bulletin, 35.

Tailleur, I.L., C.G. Mull, and H.A. Tourtelot. 1973. A skeleton in Triassic rocks in the Brooks Range foothills. *Arctic* 26:79–81.

Taylor, M.E. and A.G. Hannam. 1987. Tooth microwear and diet in the African Viverridae. *Canadian Journal of Zoology* 65:1696–1702.

Teaford, M.F. 1985. Molar microwear and diet in the genus *Cebus*. *American Journal of Physical Anthropology* 66:363–370.

Teaford, M.F. and A. Walker. 1984. Quantitative differences in dental microwear between primate species with different diets and a comment on the presumed diet of Sivapithecus. *American Journal of Physical Anthropology* 64:191–200.

Thomas, D.C. and J.E. Edmonds. 1984. Competition between caribou and muskoxen, Melville Island, NWT, Canada. *Biological Papers of the University of Alaska* 4:93–100.

Thurber, J.M. and R.O. Peterson. 1991. Changes in body size associated with range expansion in the coyote (*Canis latrans*). *Journal of Mammalogy* 72:750–755.

Townsend, V.R., Jr., J.A. Akin, B.E. Felgenhauer, J. Dauphine, and S.A. Kidder. 1999. Dentition of the Ground Skink, *Scincella lateralis* (Sauria, Scincidae). *Copeia* 1999:783–788.

Tykoski, R.S. and A.R. Fiorillo. 2013. The braincase of *Pachyrhinosaurus perotorum* compared to other Pachyrhinosaurus species, and its utility for species-level recognition. *Earth and Environmental Science Transactions of the Royal Society of Edinburgh* 103:487–499.

Tykoski, R.S. and T. Rowe. 2004. Ceratosauria. In *The Dinosauria*, eds. D.B. Weishampel, P. Dodson, and H. Osmolska, 47–70. Berkeley, CA: University of California Press.

Ungar, P.S. 1994. Incisor microwear of Sumatran Anthropoid Primates. *American Journal of Physical Anthropology* 94:339–363.

Ungar, P.S. 1998. Dental allometry, morphology, and wear as evidence for diet in fossil primates. *Evolutionary Anthropology* 7:205–217.

von Humboldt, A. 1808. *Ansichten der Natur mit wissenschaftlichen Erlauterungen*. Tübingen, Germany: J.G. Cotta.

Van Valkenburgh, B. and R.E. Molnar. 2002. Dinosaurian and mammalian predators compared. *Paleobiology* 28:527–543.

Van Valkenburgh, B. and R.K. Wayne. 1994. Shape divergence associated with size convergence in sympatric East African jackals. *Ecology* 75:1567–1581.

Walker, A. and M. Teaford. 1989. Inferences from quantitative analysis of dental microwear. *Folia Primatologia* 53:177–189.

Walker, A., H.H. Hoeck, and L. Perez. 1978. Microwear of mammalian teeth as an indicator of diet. *Science* 201:908–910.

Walker, D.A., N.A. Auerbach, J.G. Bockheim, F.S. Chapin, W. Eugster, J.Y. King, J.P. McFadden et al. 1998. Energy and trace-gas fluxes across a soil pH boundary in the Arctic. *Nature* 394:469–472.

Watanabe, A., G.M. Erickson, and P.S. Druckenmiller. 2013. An ornithomimosaurian from the upper cretaceous prince creek formation of Alaska. *Journal of Vertebrate Paleontology* 33:1169–1175.

Weems, R.E. and R.B. Blodgett. 1996. The pliosaurid *Megalneusaurus*: A newly recognized occurrence in the Upper Jurassic Naknek Formation of the Alaska Peninsula. In *Geologic Studies in Alaska by the U.S. Geological Survey, 1994*, eds. T.E. Moore and J.A. Dumoulin, 169–176. United States Geological Survey Bulletin 2152.

Weishampel, D.B. and R.E. Heinrich. 1992. Systematics of Hypsilophodontidae and basal Iguanodontia (Dinosauria: Ornithopoda). *Historical Biology* 6, 159–184.

Wilson, M.V.H. and A.M. Murray. 2008. Osteoglossomorpha: Phylogeny, biogeography, and fossil record and the significance of key African and Chinese fossil taxa. In *Fishes and the Break-up of Pangaea*, eds. L. Cavin, A. Longbottom, and M. Richter, 185–219. London, UK: Geological Society of London, Special Publication 295.

Winkler, D.A., P.A. Murry, and L.L. Jacobs. 1998. The new ornithopod dinosaur from the Proctor Lake, Texas, and the destruction of the family Hypsilophodontidae. *Journal of Vertebrate Paleontology* 18(Suppl):87A.

Witmer, L.M. and R.C. Ridgely. 2008. Structure of the brain cavity and inner ear of the centrosaurine ceratopsid dinosaur *Pachyrhinosaurus* based on CT scanning and 3D visualization. In *A New Horned Dinosaur from an Upper Cretaceous Bone Bed in Alberta*, eds. P.J. Currie, W. Langston, Jr., and D.H. Tanke, 117–144. Ottawa, Canada: NRC Research Press.

Wright, D.H. 1983. Species-energy theory: An extension of species-area theory. *Oikos* 41:496–506.

Xing, H., X. Zhao, K. Wang, D. Li, S. Chen, C.M. Jordan, Y. Zhang, and X. Xu. 2014. Comparative osteology and phylogenetic relationship of Edmontosaurus and Shantungosaurus (Dinosauria: Hadrosauridae) from the Upper Cretaceous of North America and East Asia. *Acta Geological Sinica (English Edition)* 88:1623–1652.

Xing, H., J.C. Mallon, and M.L. Currie. 2017. Supplementary cranial description of the types of *Edmontosaurus regalis* (Ornithischia: Hadrosauridae), with comments on the phylogenetics and biogeography of Hadrosaurinae. *PLoS ONE* 12(4):e0175253. https://doi.org/10.1371/journal.pone.0175253.

Zheleznoav, A. and T.M. Okuneva. 1972. Triasovaya sistema: Severo-Vostochnyy region, Kolvillskiy basseyn sedimentatisii. In *Geologiya Severo-Vostochnoy Azii, Tom 11, Stratigrafiya i Paleogografiya*, ed. V.N. Vereshchagin, 1–253. Izddanja Nedra 25. Leningrad: Leningrad Otdelenie.

Zinke, J. and O.W.M. Rauhut. 1994. Small theropods (Dinosauria, Saurischia) from the Upper Jurassic and Lower Cretaceous of the Iberian Peninsula. *Berliner geowissenschaftliche Abhundlichen* 13:163–177.

# 5 The Footprints

## INTRODUCTION

Fossil footprints and fossil bones provide different sorts of information about the animals that left them in the rock record. Bones record the bodies, and footprints record the behavior, forever preserved in stone. Footprints also provide some level of taxonomic identification because dinosaurs with differently constructed feet left behind characteristic footprints. Thus, while bones provide the opportunity to identify species and understand the skeletal movements of who might have been at the proverbial dance, tracks tell us what the dance was.

While there had been only a smattering of reports of fossil footprints around Alaska, the last 20 or so years of fieldwork in Alaska has produced a profusion of fossil vertebrate footprints (Gangloff, 1998; Fiorillo and Parrish, 2004; Fiorillo et al., 2009, 2011, 2012, 2014a,b,c; Fowell et al., 2011; Fiorillo and Adams, 2012; Tomsich et al., 2014; Fiorillo et al., 2015; Fiorillo and Tykoski, 2016; Flaig et al., 2017). This remarkable growth in the documented record of fossil footprints stems from the simple fact that one animal can make a great number of tracks over the course of its lifetime. Hence, there is a much higher probability that one would find footprints of an animal before finding its fossil bones.

The first discovery of fossil footprints and their recognition by local peoples is unknown. But since Herodotus first described an impression in Ancient Greece, or when the indigenous peoples of the Dampier Peninsula in Australia recognized dinosaur tracks and incorporated them into their oral cultural heritage, footprints in rock have been ascribed to a variety of trackmakers including humans, mythological animals, or gods (Mayor and Sarjeant, 2001; Salisbury et al., 2016). William Sarjeant was a renowned paleontologist who spent the majority of his career at the University of Saskatchewan. Though he started as a paleontologist interested in marine fossils and the history of the Earth Sciences, Sarjeant later became interested in the world of fossil footprints, and he deserves a great deal of credit for erecting improved protocols for their study. Sarjeant (1983) stated that the footprints of dinosaurs "retain a glamour in the eyes of people at large that marine trace fossils seem likely never to attain," but he also pointed out that in the second half of the 20th century, the study of fossil vertebrate footprints "ceased to be considered a respectable study by professional geologists." Sarjeant went on to note that "vertebrate paleontologists considered them to be without interest, perceiving neither that animal tracks painted a dynamic picture that bones could not match, nor that footprints provided details of the soft morphology of the foot not obtainable from bones." Sarjeant was describing a perception with which he certainly did not agree, and neither do I.

Sarjeant's reference to pre-20th century work refers largely to the seminal studies of Edward Hitchcock, a clergyman and geologist. Hitchcock's taxonomic work was the first systematic study of ichnology (e.g., Hitchcock, 1858) and his collection of tracks was by far the most extensive ever assembled at that time (Pemberton et al., 2007). Among his many contributions, Hitchcock is credited with describing a famous slab containing several small dinosaur footprints unearthed by Pliny Moody in 1802 in New England (Steinbock, 1989; Pemberton et al., 2007). After years serving as part of the doorway on a farm in South Hadley, Massachusetts, this slab was acquired by Hitchcock and described. While almost assuredly people previously had observed dinosaur footprints, the slab uncovered by Moody and described by Hitchcock is considered to be the first authenticated record of dinosaur footprints (Steinbock, 1989) ever properly recorded.

Another notable worker within this early "window of respectability" was Roland T. Bird, an avid fossil collector for the American Museum of Natural History in New York. Bird is perhaps most remembered for his collecting of fossil footprints, including the first recognized sauropod footprints in the world. He is particularly noted for his recovery of the famous Glen Rose dinosaur trackway from the Paluxy River of Texas discovered in the late 1930s. This set of tracks includes the footprints of a four-legged sauropod dinosaur, the first recognition of sauropod tracks in the world, showing an animal that was purportedly being tracked and attacked by a large predatory dinosaur, an interpretation that has since been disputed (Lockley, 1991). The contributions and historical significance of these two notable workers are extensive, and historians have summarized the work accordingly (e.g., Lesley, 1877; Hitchcock, 1895; Bird, 1985; Farlow and Lockley, 1989), while Pemberton and others (2007) go so far as to point out the work of Hitchcock and Bird established North America as the center of the field of vertebrate ichnology early on.

Hitchcock was followed by Donald Baird (1957), who pointed out in his work on the Early Mesozoic reptile tracks of New Jersey that footprints are not a simple record of the anatomy of the trackmaker. Fossil footprints also record how a foot behaves during locomotion in contact with a particular substrate. In other words, they record the relationship between an animal and its immediate environment. The recognition of that relationship within the fossil record is a watershed concept that changed the perception of the utility of tracks and opened the door to the link with value and information gained through tracking modern animals.

In his book, *A Field Guide to Animal Tracks*, Olaus Murie (1954) asked the simple question "What has happened here?" He pointed out the value and richness of tracks for understanding modern ecosystems, and that even single tracks serve as positive data in understanding what animals might be present within an ecosystem, what they were doing, and where they are going. A growing number of resources are now available on tracking modern animals (Murie, 1954; Halfpenny, 1986; Forrest, 1988; Skalski, 1991; Rezendes, 1999; Elbroch, 2003). Building off this concept, and Baird's contribution from the study of fossil tracks, Martin Lockley (1991) became one of the first to comprehensively apply the concepts and practices of modern tracking to the fossil record.

There is something evocative, for instance, in the study of modern human footprints that provides an instant connection between trackmakers and their journeys (Bennett and Morse, 2014), and while telemetry is an invaluable tool for wildlife biologists in modern tracking of animals, Paul Rezendes (1999) pointed out that tracking, the simple following of a trail of footprints, is akin to opening the door to the life of an animal. Similarly, James Halfpenny (1986) suggested that moving from track identification to tracking is to begin a detective game into the secret lives of trackmakers and their lives, into what most compels an animal—its stomach and its hormones. Tracking modern animals, as with extinct dinosaurs, involves the need to collect solid data, test hypotheses, and to research the animals and their respective habits in the habitats where the tracks were left. As Mark Elbroch (2003) put it, "tracking is the meeting place between storytelling and science," requiring rigorously collected data and the imagination to make plausible sense of it. Such perspectives on tracking modern animals suggest that the study of modern footprints is not purely empirical but rather there is an educational process, like reading, that builds experience. This experience provides an appropriate element of imagination (Liebenberg, 1990; Rezendes, 1999).

The study of trace fossils—the marks, including footprints, left in sediment by animal behaviors—was pursued by such historic giants in natural history as Leonardo da Vinci and Ulisse Aldrovandi (Baucon, 2008, 2010). While Aldrovandi observed and described invertebrate traces without knowing what they were, da Vinci is credited with understanding traces as biogenic structures and is argued to be the father of ichnology, the study of tracks and traces (Baucon, 2010). Ichnology is the science that connects paleontology and biology with sedimentology and stratigraphy, linking these disciplines in ways most body fossils cannot (Buatois and Mangano, 2011). Trace makers not only tell about the organisms but also their intimate association with the ancient environment—the sedimentary processes and features of the environment (Eiseman and Charney, 2010). The associations of species inhabiting the environment provide fundamental depth to the understanding of ancient life.

## FORMING A FOOTPRINT

Since animals have been able to move, footprints have had the potential to be preserved in the geologic record. Reconstructing how extinct animals moved and where they moved to and from is guided in part by the study of the footprints they left behind. Proper understanding of how animals move across a variety of substrates has implications beyond the biological study of a particular organism, as locomotion studies have contributed greatly to other fields such as robotics, where despite the dissimilarity in shape, arthropod locomotion is used in modeling the stance and swing of leg movements of robots (e.g., Altendorfer et al., 2002).

Traditionally, paleontologists have modeled track formation as the result of an animal walking on a layer of wet unconsolidated mud or sand that is competent, or stiff enough to hold the shape of the print without falling apart. That surface then hardens as it dries. The dried layer preserves the footprint and is more resistant to erosion than unconsolidated mud so when the next immersion under water occurs and sediment is deposited, the track is preserved and infilled by burial from a new layer of mud or sand (Figure 5.1b). With time the sediment is further dewatered and transformed into shale, mudstone, or sandstone. Over time, as erosion sets in, if the original surface on which the animal walked becomes more weather resistant than the sedimentary layer infilling the track, then the preserved track will be an impression, which represents the surface the animal walked on (Figure 5.1), but if

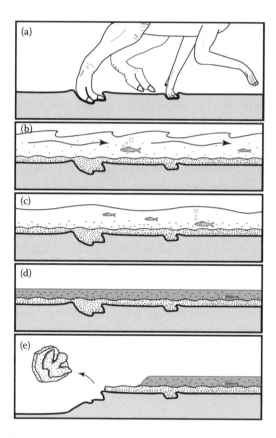

**FIGURE 5.1**  Depiction of one sequence of events that can create a footprint impression in positive relief. A dinosaur first walks across a fine sediment substrate (a). The track layer is buried by sandstone, and over time lithifies (b–d). The more resistant layer of sandstone that filled in the initial track impression erodes out of the sedimentary section and remains preserved (e). This type of track preservation represents infilling rather than the surface the animal walked on in life. In other words, these types of tracks are stratigraphically upside down.

the overlying sedimentary layer is more resistant to erosion than the underlying layer, then the original track will have served as a mold and is eroded away. The resulting track is preserved in raised relief, and these raised tracks are a cast of the original footprint (Figure 5.1e).

While this may be the manner in which many fossil vertebrate tracks are preserved, raised relief tracks can also be formed in an alternative manner. Bearing in mind that dinosaurs were often multiton animals, as a dinosaur lifts one of its feet in walking or running, the weight of the animal is redistributed to the other foot or feet. That pressure can compact the underlying sediments, creating a differential in substrate hardness such that subsequent erosion after lithification removes the less compacted surrounding sedimentary rock, leaving the compacted track impression in raised relief (Figure 5.2). In contrast to the first model where finding an infilling is actually the representation of a track turned upside down, this second model

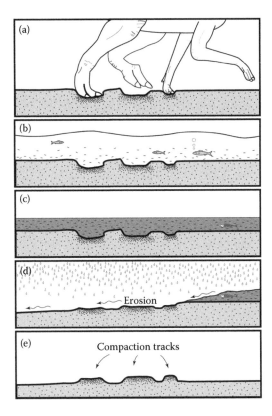

**FIGURE 5.2** Depiction of a second sequence of events that can create a footprint impression in positive relief. A dinosaur walks across a substrate, compacting the ground at the points of contact relative to the rest of the substrate (a). The original surface is buried, and over time lithifies (b, c). Erosion removes the surrounding softer matrix surrounding the original footprints, leaving the original compacted footprint features in raised relief (d, e). These raised tracks are stratigraphically right side up, and they may or may not represent the original surface of contact by the animal's foot.

produces a bas relief of a track that is right side up. The conditions for preservation of tracks pertaining to each model can be highly localized to the point that examples of both models can be found within meters of each other in a single outcrop. Careful observation is crucial to determine which way is stratigraphically up—in other words, in which direction, up or down, do the stacked layers of rocks get younger.

Anyone who has ever walked a beach or through mud has likely noticed how their footprints can vary along their walk. In some places the footprints may show great detail, while others may show much less detail such that not even toe impressions are detectable (Figure 5.3). Thus, the story of footprints is clearly much more complicated because of this sort of variation. A great deal of scientific study has been applied to this simple observation. Numerous workers have correctly pointed out the various pitfalls associated with the proper identification of fossil tracks and their makers. The quality of footprint preservation and hence track morphology can be

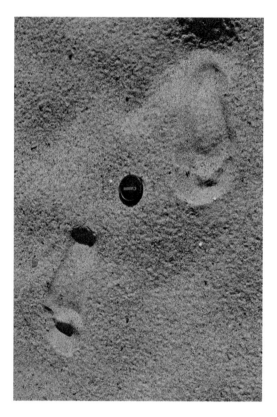

**FIGURE 5.3** Modern human footprints on a dry sandy river bar along the Charley River, Alaska. Notice the lack of definition of the tracks, particularly with respect to distinguishing individual toes.

influenced by sediment texture, consistency, and moisture (Rezendes, 1999; Elbroch and Marks, 2001; Nadon, 2001; Gatesy, 2003; Smith and Farlow, 2003; Manning, 2004; Fiorillo, 2005; Gatesy et al., 2005; Fiorillo, 2005; Farlow et al., 2006; Milàn and Bromley, 2006, 2008; Lockley, 2009; Razzolini et al., 2014; Figure 5.4). Even within known modern trackmakers, significant morphological variability can occur due to variations in sediment texture, consistency, and moisture (Rezendes, 1999; Elbroch and Marks, 2001; Fiorillo, 2005; Figure 5.4).

Ancient microbes have a role in survivability of high-quality footprints and their preservation as fossils (Kvale et al., 2001; Porada et al., 2007; Marty et al., 2009; Carvalho et al., 2013; Dai et al., 2015) because they protect exposed footprints, which can rapidly degrade due to erosion, bioturbation, and deformation during additional depositional events (Nadon, 2001) by creating a biofilm that stabilizes the surface and leads to early diagenesis (Kvale et al., 2001; Noffke et al., 2001; Noffke, 2010). Microbial mats grow in wet or moist conditions and are produced by photosynthesis, much like plants grow (Figure 5.5). The extent of these mats can be influenced by the interaction between the microbes and the clastic or chemical sedimentation rate, light, salinity, and temperature (Kendall and Skipwith, 1968; Burne and Moore,

**FIGURE 5.4**    Modern bird tracks on a riverback along the Charley River, Alaska. The moisture content and grain size on this surface were variable. As a result, the bird tracks, which were clearly left by the same bird, can be seen as tridactyl (A) and didactyl (B) tracks on the same trackway.

1987; Demicco and Hardie, 1994; Dupraz et al., 2004; Marty et al., 2009), and the mats can have variable effects on the preservation of footprints. Nonetheless, microbial mats contribute to the preservation of very well-defined footprints (Marty et al., 2009), and well-preserved tracks are crucial for ichnotaxonomic identification.

Early work on recognizing microbial mats in the fossil record centered on understanding biodiversity in the Ediacaran, approximately 635–540 million years ago. It was this work, which centered on investigating some of the earliest life on Earth, that produced the term "elephant skin" texture or structure for the wrinkled pattern produced by microbial mats (e.g., Seilacher et al., 2005). Such structure is not

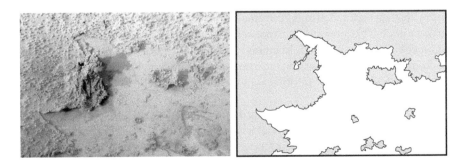

**FIGURE 5.5**    Photograph and line drawing of a modern microbial mat growing in Stony Creek, Denali National Park, Alaska. The shaded area on the line drawing corresponds to the microbial mat. Note the torn and wrinkled aspects of the mat.

**FIGURE 5.6** Comparison of a fossilized microbial mat (a) and dinosaur skin (b). Both fossils were found in the Lower Cantwell Formation of Denali National Park, Alaska. The microbial mat (specimen DENA 20918, housed at the Perot Museum of Nature and Science) exhibits the wrinkled texture referred to as "elephant skin." The dinosaur skin impression is from within a *Hadrosauropodus* track that remains in place in the park.

restricted to this window of the Precambrian, as it can be found throughout the geologic column. In the case of dinosaur-bearing rocks, such textures produced by microbial mats can appear strikingly similar to the texture produced by dinosaur skin (Figure 5.6).

Strangely enough, the step of a dinosaur that produces a track in the top layer of soft sediment can be propagated through underlying thin layers of sediment if they have the right degree of softness and stiffness to hold what are called "undertracks." The recognition of true tracks versus undertracks is of great importance to ichnology in general and especially the naming of tracks, ichnotaxonomy. True surface tracks are recognized by the presence of skin impressions, little ridges (called displacement rims) of sediment around the edges, messy tracks with distorted outlines made in fluid top layers, detached mud clasts, or collapse-and-flow features like miniature landslides inside the track (Manning, 2004).

Undertracks, preserved below the actual surface upon with the animal walked, lack these features, but they are important to recognize given that many dinosaurs were multiton animals. The weight of these large animals means that during a step on the ground, not only is the actual surface compressed, but subjacent horizons below that surface can also be deformed from the pressure generated by the track-maker's foot. This pressure is expressed downward and radially outward into the sediment (e.g., Gatesy, 2003; Manning, 2004; Milàn and Bromley, 2006). Because undertracks are already buried at the time of formation, the potential for their preservation is high, but their information is reduced because they preserve a distorted view of the original print.

The formation of undertracks formed in competent sediment—meaning the substrate could withstand the pressure of an animal moving across it without the surface rupturing—is one thing, but their formation in underlying sediments not specifically formed by gross weight deformation is another. Gatesy (2003) examined the relationship of dinosaur feet to the substrate by examining incompetent—but not completely incompetent—surfaces. He noted that toes can enter the substrate splayed. As weight shifts during a step and toes slice through the sediments, the toes adduct and

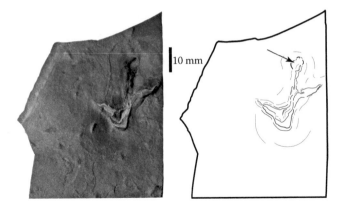

**FIGURE 5.7** Small fossil bird track from the Lower Cantwell Formation (Cretaceous) of Denali National Park, Alaska (specimen DENA 20948, housed at the Perot Museum of Nature and Science). Notice the feature highlighted by the arrow. This feature is the result of the substrate being very soft such that the foot moved through the subsurface during the step cycle and was drawn through the middle digit as the bird prepared for its next step by this foot.

flex as they are lifted. Often the foot leaves the subsurface through the initial impression made by the middle digit, creating an enlarged and somewhat circular feature in that part of the trace (Gatesy, 2003; Gatesy et al., 2005; Falkingham and Gatesy, 2014; Figure 5.7). The resulting tracks, though formed in layers below the surface, are real tracks formed at the time of motion by actions of the foot in direct contact with sediment, as opposed to undertracks, with which the foot has no contact. And while the focus on this category of track has been on larger, non-avian dinosaur footprints, even smaller animal tracks, such as those of small fossil birds, can exhibit this type of formation, indicating that body mass is not factor in this style of track formation (Figure 5.7).

## CHARISMATIC PALEOFAUNA

"Charismatic megafauna" is a favorite term for wildlife managers working in United States National Park units, referring to the large animals within an ecosystem that have widespread popular appeal, such as Gray wolves (*Canis lupus*), Grizzly bears (*Ursus arctos*), Moose (*Alces alces*), Caribou (*Rangifer tarandus*), and Dall sheep (*Ovis dalli*). Though there is no formal definition for charismatic megafauna, the term carries weight within conservation biology circles, as its application to specific wildlife can lead to funding for research and/or protection (e.g., Sergio et al., 2006; Ducarme et al., 2013).

Dinosaurs have tremendous popular appeal, with museums registering major boosts in attendance with every traveling dinosaur exhibit or renovated dinosaur hall. While one might think that this appeal is a fairly recent phenomenon, in fact the public seems to have loved dinosaurs since the very first dinosaur displays. One of the first dinosaur skeletons ever mounted was Benjamin Waterhouse Hawkins'

*Hadrosaurus foulkii* at the Academy of Natural Sciences of Philadelphia in 1868. The public response to the *Hadrosaurus foulkii* skeletal exhibit was so overwhelming as to cause chaos in what had been a conforming museum setting. The curators lobbied for an admission fee to the museum, for the first time in the history of the institution, as means of controlling the enthusiastic crowds (Peck, 2012). Based on the historical popular appeal of dinosaurs, one could refer to dinosaurs, known from either fossil bones or footprints, as "charismatic paleofauna," a term that has informally gained some acceptance within U.S. National Park Alaska Region parks.

Prior to 2001, there were no records of dinosaurs from any National Park in Alaska. Given the high profile of paleontology in the public eye, starting in 1999, funding was provided for exploratory work on the paleontological resources within national parks in the state. As a result of this initiative, there are now four Alaskan parks that have a Cretaceous dinosaur record.

The first park to produce a recognized record is Aniakchak National Monument, where tracks were attributed to hadrosaurs (Fiorillo and Parrish, 2004). For most people, the mention of Aniakchak National Monument is likely to lead to a search of the map of Alaska in an effort to determine the location of this park. It is so isolated that the National Park Service adds to its tourism information the tag line "no lines, no waiting." By Alaska standards it is a very small park, only approximately 2400 square kilometers in size, an area about one and a half times the size of the island of Oahu in the Hawaiian island chain. Located along the Alaska Peninsula, it is along the northern Alaska portion of the "Ring of Fire," the volcanically active geologic boundary around the Pacific Basin. The most prominent geographic feature of the park is the Aniakchak Caldera, a 10-kilometer wide, almost 800-meter-deep feature that resulted from the collapse of a 2100-meter-tall volcano over 3600 years ago (Pearce et al., 2004).

The most recent volcanic activity in Aniakchak occurred in 1931, the aftermath of which was documented by the explorer, geologist, and Jesuit priest Father Bernard Hubbard. Father Hubbard, nicknamed the Glacier Priest, was asked to explore Aniakchak shortly after the 1931 eruption because he had conducted the first serious geologic exploration of the area the previous year, shortly before the eruption. Hubbard was a tireless explorer and an entertaining public speaker who provided memorable perspectives on the need to gather data in the field. In his words, "A lost explorer is a contradiction in terms after all. To be more correct, an explorer is lost all the time, otherwise he would not be exploring" (Hubbard, 1943, p. 42).

Given the more dramatic and recent volcanic history of the park which drew such colorful attention, it seems improbable that this first dinosaur record for the Alaska Region National Park Service would have come from this place. The initial dinosaur track discovery at Aniakchak occurred on a whitewater raft trip down the Aniakchak River starting in the Aniakchak Caldera and ending in Aniakchak Bay. That first record consisted of a series of three dinosaur tracks on a large talus block fallen from a cliff face of Chignik Formation, an Upper Cretaceous rock unit that crops out along the Alaska Peninsula (Fiorillo and Parrish, 2004). These first tracks are on a light- to medium-gray, fine-grained sandstone block (Fiorillo and Parrish, 2004). The tracks comprise a single large tridactyl, or three-toed, impression, where the toes are long, broad, and rounded, and two smaller, crescent-shaped, associated impressions, interpreted as a foot (pes) and two hand (manus) prints, respectively. Other pes tracks

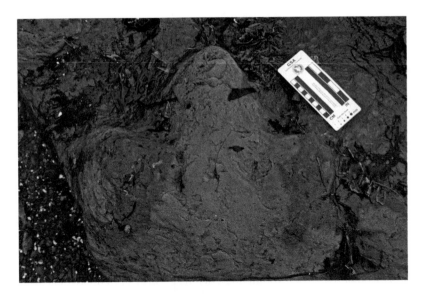

**FIGURE 5.8** Tridactyl track from the Chignik Formation (Cretaceous) of Aniakchak National Monument, Alaska. The trackmaker for this track was likely a hadrosaur.

similar in morphology to those of the initial discovery have come to light from the same outcrop belt within Aniakchak National Monument (Figure 5.8; Fiorillo et al., 2004). The tracks were made by a large ornithopod dinosaur. Based on the occurrence of the bivalves *Inoceramus balticus* var. *kunimienis* and *I. schmidti*, as well as the ammonite *Canadoceras newberryanum*, the Chignik Formation is considered to be late Campanian–early Maastrichtian (Detterman et al., 1996). Given this age then, the age of the Chignik ornithopod was most likely a hadrosaur (Fiorillo and Parrish, 2004).

Other Alaskan parks having dinosaur footprint records include the enormous Wrangell–St. Elias National Park where ornithopod and small theropod tracks were found (Fiorillo et al., 2012), and the somewhat smaller but still huge Yukon–Charley Rivers National Preserve where hadrosaur tracks were found (Fiorillo et al., 2014b). Wrangell–St. Elias National Park is located in southeastern Alaska. It is the largest national park in the United States at a whopping 54,000 square kilometers, approximately the size of the states of Connecticut, Massachusetts, and Rhode Island combined. Aside from the size of the park, and though a land perhaps best known for mountains grown from the fire of volcanism or the ice that now covers much of the park, what is now Wrangell–St. Elias National Park is historically significant in terms of the development of natural history studies in North America. It was the almost 5500-meter-tall Mount St. Elias within this park that in 1741 provided Vitus Bering and his crew their first sighting of Alaska. A small field party that included the naturalist Georg Wilhelm Steller landed on the coast. Steller was the first European to describe North American fauna and flora along the coast outside of what is now the park. His documentation of the bird now known as the Steller's Jay was used as evidence to argue that Russia and Alaska were separate land masses.

The land within the boundaries of Wrangell–St. Elias National Park was traversed in the mid-1880s by Lieutenant Henry T. Allen of the United States Army and his party of three others on their way into the Alaska Interior. Allen's exploration up the Copper River and into the Alaska Interior opened the way for the white settlement. His expedition is one of the most important in the history of Alaskan exploration. Allen's group was greatly aided by the support of local native communities, and in particular the individual Nicolai, an Ahtna chief. While at first mistaking Allen's group for Russians, to whom the Ahtna were hostile, once Nicolai realized they were Americans, he provided the much-needed guidance for the mapping party up the Copper River drainage, where Allen's group eventually left the drainage, crossed the Alaska Range, and entered the interior part of the state. One result of Allen's expedition was that this group became the first to map this terrain.

In 2010, outside of the old mining town of Chisana in the Nutzotin Basin, our paleontological reconnaissance (Figure 5.9) discovered a small assemblage of tracks. A single small pes impression with three long, thin digit impressions and a curved, sinusoidal middle digit is diagnostic of some smaller theropod tracks (Fiorillo et al., 2012). The other track from within this assemblage, also a pes impression, has three broad, clearly defined digit impressions with blunt ends lacking claws. Based on the shape, this impression was attributed to an ornithopod, possibly a hadrosaur (Fiorillo et al., 2012).

These tracks were preserved in a depositional setting very different from tracks found elsewhere in Alaska. Within these Nutzotin Basin sedimentary rocks, which

**FIGURE 5.9** View of the landscape of Wrangell–St. Elias National Park, Alaska. The author (left) and Yoshitsugu Kobayashi (Hokkaido University Museum, right) for scale. (Photograph courtesy of Linda Stromquist, National Park Service [retired].)

are still unnamed, there is an abundance of very coarse-grained material—conglomerates and very coarse sandstones. The abundance of these coarse-grained rocks suggests that the area was being drained by high-energy river systems. Further, the presence of upright trees with their root systems intact suggests that there were high rates of sediment aggradation out on the floodplains (i.e., frequent floods). Within these unnamed rocks there is also an abundance of charcoal, which suggests that fire was prevalent within this ancient ecosystem. So, unlike other National Park areas where dinosaur tracks have been found, within this basin deposition in the paleoecosystem was dynamic, where high-energy rivers flowed that often breached the levees and overflowed onto the floodplains, and the forests were prone to fire disturbance (Fiorillo et al., 2012).

Yukon–Charley Rivers National Preserve is located in the Alaskan Interior. It is much smaller than Wrangell–St. Elias, coming in at a little more than a mere 10,000 square kilometers. The Yukon River, the main feature of the park, is choked with silt, but its tributaries run clear, because this part of Alaska was not glaciated during the Pleistocene and therefore hills drained by the tributaries do not flow through or transport the ground-up rock flour and particles gouged by glaciers. Yukon–Charley Rivers National Preserve lies in a region well known for gold mining, particularly within the rock unit that produced the first record of a dinosaur in this part of the state (Fiorillo et al., 2014b).

The outcrop pattern of Upper Cretaceous rocks is limited by dense, forested, vegetative cover. Even by Alaskan standards, the outcrops are few and far between, very difficult to reach because of the rugged landscape, and are mostly restricted to the banks of tributaries of the Yukon River (Figure 5.10). It took, for example, four field

**FIGURE 5.10** Example of typical outcrop pattern for Cretaceous–Tertiary rocks of Yukon–Charley Rivers National Preserve. Federico Fanti (University of Bologna) and Stephen Hasiotis (University of Kansas) for scale.

expeditions to discover the rock exposure that produced the park's dinosaur record, and a fifth expedition to the outcrop to further sample and confirm the findings. Based on size and shape, the morphology of the tracks is consistent with that of tracks made by hadrosaurs (Fiorillo et al., 2014b).

The dinosaur records from these three Alaskan parks, though exciting and new, are of a limited nature; the track record is not particularly robust compared to places like the well-studied Mesozoic rocks of the Colorado Plateau in western North America. Despite that lack of abundance, what makes these Alaskan discoveries important is that they are of similar geologic age—that is, latest Cretaceous—and include a window of approximately 70–66 million years ago. Therefore these discoveries contribute to the evidence of the existence of a high-latitude terrestrial ecosystem with large-bodied dinosaurs that was widespread across northwest North America during the Cretaceous.

By far, the park in Alaska that has produced the most remarkable records of fossil footprints of this age is Denali National Park in the central Alaska Range. The suite of footprint fossils found there rival those anywhere on the globe in terms of abundance and quality of preservation. The track record of Denali anchors the ancient vertebrate story of the high latitudes as told by footprints (Figure 5.11).

Denali National Park is famous for its wildlife and is home to the tallest mountain in North America, formerly known as Mount McKinley and more recently officially known as Denali. Within the almost 25,000 square kilometers of Denali National Park, and in contrast to the other parks, we now have thousands of known dinosaur footprints, a fact that is all the more remarkable given that prior to 2005 none had been recognized. As a result of more than 10 years of fieldwork in Denali, there are now some fifteen different vertebrate ichnotaxa recorded from the Lower Cantwell Formation of Denali National Park (Table 5.1).

The first discovery of a footprint occurred in 2005 when Dr. Paul McCarthy, a professor of geology at the University of Alaska in Fairbanks, took two of his

**FIGURE 5.11** Illustration of the variety of Cretaceous vertebrate animals now recorded from the Lower Cantwell Formation of Denali National Park, Alaska. (Artwork by Karen Carr. The original artwork is on display at the Murie Science and Learning Center in Denali National Park.)

## TABLE 5.1
## List of Fossil Vertebrate Ichnogenera from the Lower Cantwell Formation of Denali National Park, Central Alaska Range, Alaska

Fish
    *Undichna* isp.
Reptilia
    Pterosauria
        Pteraichnidae
            *Pteraichnus* isp. (large form)
            *Pteraichnus* isp. (small form)
Saurischia
    Theropoda
        *Eubrontes* isp.
        *Saurexallopus* isp.
        Dromaeopodidae
            *Menglongipus* isp.
        Aves
            *Ignotornis mcconnelli*
            *Aquatilavipes swiboldae*
            *Magnoavipes denaliensis*
            *Magnoavipes* isp.
            *Gruipeda vegrandiunus*
            *Uhangrichnus chuni*
            *Uhangrichnus* isp.
Ornithischia
    Hadrosauridae
        *Hadrosauropodus* isp.
    Ceratopsidae
        *Ceratopsipes* isp.

*Source:* With the exception of *Ceratopsipes* isp., these ichnotaxa have been described earlier (Adapted from Fiorillo, A.R. and T.L. Adams. 2012. *Palaios* 27:395–400; Fiorillo, A.R. et al. 2009. *Palaios* 24:466–472; Fiorillo, A.R. et al. 2011. *Journal of Systematic Palaeontology* 9:33–49; Fiorillo, A.R. et al. 2014a. Theropod tracks from the Lower Cantwell Formation (Upper Cretaceous) of Denali National Park, Alaska, USA with comments on theropod diversity in an ancient, high-latitude terrestrial ecosystem. In *Tracking Dinosaurs and Other Tetrapods in North America*, eds. M. Lockley and S.G. Lucas, 429–439. Albuquerque: New Mexico Museum of Natural History and Science; Fiorillo, A.R. et al. 2014c. *Geology* 42:719–722.) The attribution of a ceratopsian ichnotaxon to this fauna was made here. See text for detailed discussion of these taxa.

students to an outcrop within the park not far from the road. The group had worked hard the previous day over the course of the department's field camp to learn field methods in geology, and the students seemed to need an easier load that particular day. Paul chose an easily accessible outcrop to give an impromptu presentation about the sedimentology of the Lower Cantwell Formation. During the course of the presentation, he mentioned that there is an expectation that with some effort, dinosaur footprints might be found. The two students, Carla (Susi) Tomisch and Jeremiah Drewel, inquired as to what such a track might look like, and then pointed to a solitary, three-toed track. That track is now identified as belonging to the ichnotaxon *Eubrontes* (Fiorillo et al., 2014a). Because in 2005 it was unclear whether the newly discovered track was a unique occurrence and because Denali Park staff recognized the importance of a dinosaur discovery in Alaska's most visited national park, the decision was made by the park to support the excavation of the original track, and it has now been viewed by hundreds of thousands of visitors to the park.

Subsequent work demonstrated the presence of thousands of footprints in the Lower Cantwell Formation exposed in Denali Park. The most taxonomically diverse group of dinosaurs represented within the enormous assemblage of tracks is theropods, including Aves—birds (Fiorillo et al., 2011, 2014a; Fiorillo and Adams, 2012). There are both three-toed small theropod tracks, attributed to *Eubrontes* isp., and didactyl tracks showing only two toes, attributed to *Menglongipus* isp. (Fiorillo et al., 2014a). The latter ichnotaxon is a member of the Dromaeopodidae, the footprint taxonomic group that is the equivalent of the derived group, the Dromaeosauridae, known from skeletal remains. The Dromaeosauridae are those theropods with an enlarged claw on the second toe of their feet; thus tracks attributable to the Dromaeopodidae are two-toed to accommodate the clawed second toe being held off the ground. Thus, both primitive and more derived small theropod footprints are recorded in the Cantwell Formation.

Two-toed tracks, made by animals with the derived claw of digit II, likely reflect a grappler/slasher or scansorial hunting style by the trackmaker, which may have included a social component (i.e., pack hunting) as well, thus hunting prey animals larger than themselves (sensu Van Valkenburgh and Molnar, 2002; Manning et al., 2006, 2009; Fowler et al., 2011). In contrast, the three-toed tracks were made by predatory dinosaurs thought to have hunted prey much smaller in relative body size compared to the didactyl theropods (Van Valkenburgh and Molnar, 2002). These two foot structures, tridactyl and didactyl tracks, suggest different hunting strategies among theropods and likely niche separation within the lower Cantwell ecosystem through the process of food acquisition (Fiorillo and others, 2014a).

The Lower Cantwell Formation contains the only track evidence in northern North America of therizinosaurs (Fiorillo and Adams, 2012), another type of theropod dinosaur. The tracks left by these dinosaurs are distinctive in that the tracks are four-toed, with each digit substantially longer than wide, and the Denali tracks were attributed to the ichnogenus *Saurexallopus* (Fiorillo and Adams, 2012). Digits II and III tend to be the longest in these Alaskan tracks.

Therizinosaurs (Figure 5.12) are highly unusual theropods in that may have been herbivorous (Clark et al., 2004), which is unusual for non-avian theropods. The first of these tracks was discovered by an educator, David Tomeo, in 2010. David works at the Murie Science and Learning Center in Denali National Park. The center is devoted to promoting science and stewardship within the park, which includes telling the dinosaur stories learned from the Lower Cantwell Formation. David came across a four-toed track during one of his educational hikes. Knowing that my group would arrive later in the summer, he took a picture of this unusual track and sent it to me. My field plans for that summer shifted to include the area where David had been so we could evaluate his discovery.

**FIGURE 5.12**  Illustration of a therizinosaur, an unusual theropod dinosaur, now known from the Lower Cantwell Formation (Cretaceous) of Denali National Park, Alaska (Fiorillo and Adams, 2012). (Artwork by Karen Carr.)

Subsequently, we discovered dozens of similar therizinosaur tracks in that part of the park. Therizinosaur body fossils have been found in both Asia and North America, but none were known this far north (Fiorillo and Adams, 2012). To explain the prior known distribution of therizinosaurs in Asia and North America, it was suggested that therizinosaurs used the land bridge connection between these two continents (Zanno, 2010). The discovery of tracks attributable to therizinosaurs in Denali National Park is evidence in exactly the right place to support this biogeographic model (Fiorillo and Adams, 2012).

More than this quirk of geography, the therizinosaurs of Denali National Park offer an opportunity to speculate about the paleobiology of the group. Modern terrestrial high-latitude environments are unique in the physical constraints they impose on organisms, including marked seasonal changes in temperature, exceedingly short growing seasons, and profound shifts in the annual light regime. The presence of therizinosaurs in such a setting indicates that this seemingly enigmatic group of theropods was well adapted to the rigors of the ancient Arctic physical environment.

The Lower Cantwell Formation of Denali National Park also has an unparalleled record of fossil avian biodiversity as represented by tracks and traces, which represents the farthest north Cretaceous avian fauna known (Fiorillo et al., 2011). Tracks attributable to birds are found in multiple locations along an almost 50-km transect within the park. Some bird tracks are found with non-avian dinosaur tracks, but other surfaces with bird tracks were preserved in strata interbedded with those containing only non-avian dinosaur track-bearing rock layers.

The bird footprints are assigned to several ichnotaxa that have been found in Asia, North America, or both. The distinctions between these ichnotaxa are based on the presence of a hallux, size, shape, and angle of divarication of the digits: *Aquatilavipes swiboldae*, *Ignotornis mcconnelli*, *Uhangrichnus chuni*, *Uhangrichnus* isp., *Magnoavipes* isp., *Magnoavipes denaliensis*, and *Gruipeda vegrandiunus* (Fiorillo et al., 2011; Table 5.1). The first five ichnogenera have been reported from both Asia and North America, while the last two are ichnotaxa unique to the Lower Cantwell Formation (Fiorillo et al., 2011). Based on the size range of these avian tracks, the body sizes of these fossil birds recorded within the Lower Cantwell Formation show a size range from sparrow-sized birds such as *Gruipeda vegrandiunus* to crane-sized birds (~25%–30% larger than the modern Sandhill crane, *Grus canadensis*) such as *Magnoavipes denaliensis*.

In some cases, near-circular to sub-oval depressions have been associated with *Ignotornis* and *Aquatilavipes* tracks (Fiorillo et al., 2011). Many of these depressions are puncture-like; that is, they appear as short, sharply tapering, somewhat conical features. In some examples, the depressions are filled with a slightly coarser matrix than what is found on the surface containing the tracks, and these features indicate a stage of sedimentary infilling separate from when the track layer was laid down. As these features resemble the depressions left by the bills of modern members of the Charadriiformes as the birds probe the substrate in search for food, these fossil structures have been interpreted as similar bill probe marks recording feeding behavior by these fossil birds (Fiorillo et al., 2011).

Fossil bird tracks provide information about the ecological dynamics of the ancient Arctic. The migratory paths of some modern birds, such as Canada geese (*Branta canadensis*) and Sandhill cranes (*Grus canadensis*), in Alaska include either Asia or North America (Irving, 1960, 1972) and these birds, as well as the many others, nest in modern Alaska to take advantage of the seasonal productivity of the summer Arctic ecosystem. That the Lower Cantwell Formation records an assemblage of fossil bird tracks previously described in Asia, North America, or both, suggests that the avian fauna of Cretaceous Alaska similarly took advantage of the seasonal productivity of the ancient summer Arctic ecosystem.

While the component of the known track record in Denali National Park with the greatest taxonomic diversity is that of theropods (including birds), the most commonly found tracks in the Lower Cantwell Formation are those of hadrosaurs, the duck-billed dinosaurs. Hadrosaur tracks are wider than long, tridactyl, with digits that terminate bluntly, and have a wide, bilobed heel, all of which are key characteristics for determining the hadrosaur origin of the tracks (Currie et al., 2003; Lockley et al., 2004). Very well-preserved tracks at Denali not only have these features but also preserve skin impressions that match tubercle patterns previously identified in hadrosaurids (Lockley et al., 2004). Tracks attributable to hadrosaurs are referred to as *Hadrosauropodus* (Lockley et al., 2004), and this ichnogenus name that has been applied to the hadrosaur tracks of the Lower Cantwell Formation (Fiorillo et al., 2014c). Approximately 73% of all tracks attributable to herbivorous dinosaurs that have been found in the park so far were made by hadrosaurs.

Juvenile *Hadrosauropodus*, as well as adults, are common in Denali National Park. One tracksite, which overlooks the Toklat River, contains thousands of well-preserved hadrosaur tracks of a size range from fully grown adults to juveniles, and most of the tracks contain skin impressions (Fiorillo et al., 2014c). Track

**FIGURE 5.13**  Illustration of a herd of plant-eating hadrosaurs moving across the Cretaceous landscape of what is now Denali National Park, Alaska. This artwork represents the first effort to portray the remarkable fossil record known from this park. (Artwork by Hisashi Masuda.)

density at this site is so great that individual trackways that would have illustrated movements of individual hadrosaurs are obscured. A statistical analysis of the length and width measurements of the tracks shows a hadrosaur herd containing adults, subadults, juveniles, and very young individuals. This tracksite indicates that Arctic hadrosaurs lived in multigenerational herds (Figure 5.13), because dinosaurs lay eggs and it is unlikely that they had more than one breeding season due to the extreme Arctic climate. The smallest prints represent a current year's births, the next larger suite a previous year's births, and so on. This is a life-history pattern not previously recognized from either bonebeds or from other track assemblages (Fiorillo et al., 2014c).

While a story of parental care in polar hadrosaurs is interesting based on one site, it becomes even more exciting when it is substantiated by the discovery of additional sites that repeat the results. A second site found in the Lower Cantwell Formation also contains an abundance of *Hadrosauropodus* tracks. This second site, which is in the Big Creek drainage, an area tens of kilometers east of the first site near the Toklat River, contains a similar size range of tracks as the previous, but the preservation is much poorer due to the dinosaurs walking on coarser sediment (Figure 5.14). A quick scan of the size range of these footprints shows adults to juveniles (Figure 5.15), which not only corroborates the previous discovery, but also strongly indicates that not only did duck-billed dinosaurs live in the ancient Arctic,

**FIGURE 5.14** Photograph of a very small three-toed hadrosaur footprint from the Lower Cantwell Formation (Cretaceous) of Denali National Park, Alaska. The size of this track illustrates that hadrosaurs of all ages lived in the region during the Cretaceous.

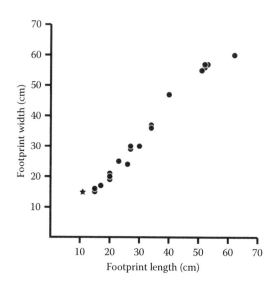

**FIGURE 5.15** Graph of length–width measurements of hadrosaur tracks from Lower Cantwell Formation exposure in Big Creek, Denali National Park. The star represents the very small track discussed in the text. Note the range in size of hadrosaur individuals from very young to fully grown adults.

but they thrived and possibly bred there. It is tempting to speculate further that the coexistence of adult, subadult, juveniles, and very young individuals in the northern polar region exhibited herd behavior, and therefore postnatal parental care, that extended beyond the nest.

Not only does this abundance of tracks tell us that hadrosaurs were the dominant large plant-eating dinosaur walking this landscape, but an additional track discovery at this second large tracksite informs us about the way the juvenile individuals moved. The posture of hadrosaurid dinosaurs has been discussed through the years (Osborn, 1912; Galton, 1970), and it is now realized that adult forms were facultative bipeds (Currie et al., 1991; Dilkes, 2001; Senter, 2012). With respect to juvenile hadrosaurs, in his detailed study of the ontogenetic growth of forelimb and hindlimb elements in the hadrosaurid *Maiasaura peeblesorum*, Dilkes (2001) suggested that juvenile forms were largely bipedal. Numerous tracks attributable to juvenile hadrosaurs have been documented from the Lower Cantwell Formation of Denali National Park (Fiorillo et al., 2014c; Fiorillo and Tykoski, 2016), including a track set from the second site discussed that is attributable to a very small hadrosaur (Fiorillo and Tykoski, 2016), and this track set adds to the discussion.

The standard equation to determine hip height from footprints of dinosaurs is approximately four times the track length (sensu Alexander, 1976; Henderson, 2003). This smallest recorded juvenile *Hadrosauropodus* pes track from the Lower Cantwell Formation is 11 cm long, providing a hip height of approximately 44 cm for the trackmaker (Fiorillo and Tykoski, 2016; Figure 5.16). Typically, the adult forms of hadrosaurs from the Campanian and Maastrichtian of North America reached body lengths of 1200 cm. Reconstruction of the hip height for the Maastrichtian hadrosaurid *Edmontosaurus regalis* is approximately 266 cm with a body length of approximately 1000 cm, providing a relationship of body length as approximately 3.76 × hip height (Henderson, 2003).

The hip height estimate from this small *Hadrosauropodus* pes impression is approximately 44 cm, yielding a body length of 165 cm (Figure 5.16). This very small *Hadrosauropodus* track set is part of a larger bedding plane that contains numerous hadrosaur tracks of various sizes that record the actions of a multigenerational herd

**FIGURE 5.16** Depiction of the body size relationship between an adult hadrosaur and the smallest hadrosaur found at the tracksite in the Big Creek drainage of Denali National Park, Alaska.

of these dinosaurs (Fiorillo and Tykoski, 2016). Adult-sized tracks within this entire assemblage are approximately 64 cm long, a length that provides a hip height for adult hadrosaurs of 256 cm (Fiorillo and Tykoski, 2016; Figure 5.16). The discovery of this set of tracks attributable to such a small individual shows that at least some juvenile hadrosaurs were facultative quadrupeds rather than obligatory bipeds, regardless of their small body size. This suggests large size, and by implication body mass was not the driving factor behind the adoption of quadrupedality in hadrosaurids.

Related to this abundance of hadrosaur tracks, but occurring outside of Denali National Park, is the recent recognition of hadrosaur tracks from the Prince Creek Formation of the North Slope of Alaska (Flaig et al., 2017). The range of sizes of the tracks from this North Slope site, which is comparable to these two tracksites in Denali National Park, is additional evidence of multiple generations and sizes of individuals that traveled together, but this time on the Arctic Alaska coastal plain of far northern Alaska. Further, by including tracks of the same size as those preserved in Denali National Park, this site provides important evidence for fully adult hadrosaurs from the Prince Creek Formation. This new record for the presence for adults provides ichnological evidence that illustrates how young and small the hadrosaur individuals are that are known from the bonebeds along the Colville River, like the Liscomb Bonebed, how much growth those young dinosaurs still would have gone through to reach adulthood, and how erecting a new taxon for these hadrosaur from the bone record is likely premature (Chapter 4).

Tracks attributable to ceratopsians, the horned dinosaurs, are generally uncommon in the rock record. Based largely on large quadruped dinosaur tracks found in the Late Cretaceous of Colorado, Lockley and Hunt (1995) erected the ichnotaxon *Ceratopsipes goldenensis*. These tracks differ from other quadrupedal dinosaur tracks, such as those made by the armored ankylosaurs, by having very blunt digit impressions. Further, these digits are of similar dimension, with no digit being most prominent (McCrea et al., 2001). The original distribution of *Ceratopsipes* in Upper Cretaceous rocks in western North America includes Alberta, Colorado, Utah, and possibly Wyoming (Lockley and Hunt, 1995), but tracks attributable to horned dinosaurs, though not ascribed to this ichnotaxon, were found in older rocks in Alaska and suggest that shortly after the Beringian land bridge formed some 110 million years ago, neoceratopsians were among the first to exploit it (Fiorillo et al., 2010).

While skeletal evidence suggested to some workers that ceratopsians held their forelimbs in a sprawled manner (Lull, 1933; Farlow and Dodson, 1974; Lehman, 1989), the track record in the lower latitudes of western North America suggests that these animals held their limbs erect (Lockley and Hunt, 1995). To resolve this difference in interpretation, Lockley and Hunt (1995) suggested that perhaps the sprawling posture was used during feeding, while the more erect posture was used during locomotion.

Within the Lower Cantwell Formation of Denali National Park, we have found several isolated occurrences of large tracks that have four or five blunt digit impressions (Figure 5.17), as well as some smaller tracks that are more arcuate in shape with a flattened or concave heel. These tracks compare favorably with *Ceratopsipes*, with the larger tracks being pes tracks and the smaller being manus tracks, though Denali pes tracks do not have as well developed a heel as *Ceratopsipes goldenensis*.

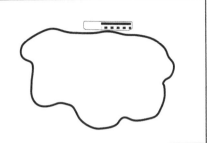

**FIGURE 5.17** Photograph and line drawing of an isolated ceratopsian (*Ceratopsipes* isp.) pes track from the Lower Cantwell Formation, Denali National Park, Alaska.

In addition, the pes tracks from Denali often have a fifth toe impression. These tracks in the Lower Cantwell Formation of Denali National Park have been found on isolated blocks, or on exposures with bedding planes of very limited extent. Typically, the tracks occur as individual tracks, and occasionally there are bedding surfaces that contain two tracks. Very rarely do the bedding surfaces contain more than two tracks. So, the size and shape of these tracks confirms the presence of ceratopsians in this high-latitude ichnofauna, but the occurrence of tracks in isolation does not help contribute to stories about stance.

## UP IN THE AIR

The Lower Cantwell Formation of Denali National Park has an abundant and somewhat diverse record of tracks made by pterosaurs, the winged reptiles often depicted flying about the heads of dinosaurs (Figure 5.18). Pterosaurs were animals that do not have modern analogs, and aspects of their paleobiology are topics of extensive discussion. For example, while most agree that the anatomy of pterosaurs is unique and traditionally interpreted as these animals primarily flew, some workers have

**FIGURE 5.18** Illustration of a large pterosaur soaring over ancient Denali National Park, Alaska. The size of the animal is based on large footprints that have been found in the Lower Cantwell Formation (Cretaceous) within the park. (Artwork by James Havens.)

suggested instead that the largest pterosaurs were terrestrial stalkers (Witton and Naish, 2008). Regardless, the track record now known in the far north does offer insight into the paleobiology of this very successful group of fossil reptiles.

Pterosaur tracks are easily identifiable because of the skeletal structure of their feet and hands, the latter typically being longer than the former. The skeletal structure of the hands shows that digit four, or the outermost digit, is highly elongated. This results in the digit being held up and off the ground, resulting in a manus impression of only the first three digits (Figure 5.19; Lockley et al., 1995; Mazin et al., 2003; Fiorillo et al., 2009, 2015). Pes tracks tend to be longer than wide, plantigrade, four-toed, and sub-triangular in shape (Lockley et al., 1995).

While tracks made by the manus may seem to be counterintuitive for animals with wings, Mazin et al. (2003) pointed out two important variables that are unique to pterosaurs. The first is that the center of mass in pterosaurs is anteriorly displaced, so that they put more body weight on their front limbs than their back ones when on the ground. Second, related to weight distribution favoring the front limbs, the surface area of the hands in pterosaurs in contact with the ground is decidedly smaller than the equivalent surface area of the feet. The increased weight over a smaller surface area would tend to yield deeper and likely more recognizable manus tracks. As a result, there is a higher probability that manus tracks rather than pes tracks would be preserved.

Only manus impressions are known from Denali National Park (Figure 5.19), and all are identified as the ichnogenus *Pteraichnus*. There is some diversity in this ichnogenus, as larger and smaller forms are both present, with the former tracks being approximately 18 centimeters long and the smaller tracks being approximately 6

**FIGURE 5.19**   Photograph of small pterosaur track found in the Lower Cantwell Formation (Cretaceous) of Denali National Park, Alaska.

centimeters long (Fiorillo et al., 2009, 2015). The tracks are tridactyl, asymmetrical, and digitigrade—that is, they are made by the digits rather than the palm. The third digit for each track is oriented posteriorly and is the longest of the digit impressions. For a sense of scale, the largest pterosaur, *Quetzalcoatlus*, had a wingspan of approximately 11 meters, and was of the size to make manus tracks appropriate to these Denali pterosaur tracks. But given the lack of skeletal confirmation, these Denali tracks do not indicate the presence of *Quetzalcoatlus* per se in the skies of this ancient high-latitude ecosystem.

Pterosaurs are considered to be warm-blooded because flying is energetically demanding. Endothermy, or internally controlled constant temperature, has been suggested for some non-avian dinosaurs as well, and is definitively present in birds, the avian dinosaurs. Polar dinosaurs in Alaska are congruent with this idea. The broad group Archosauria contains the Crocodylia, Pterosauria, and Dinosauria. Dinosaurs are present in the Cretaceous Arctic but some of their archosaur relatives are not. The distribution of fossil crocodiles in the Late Cretaceous shows no record in the ancient high latitudes (Markwick, 1998), although they are present during the Eocene, some 25 million years after most of the dinosaurs discussed throughout these chapters. This implies that, like modern crocodiles, Cretaceous crocodiles were ectothermic and not tolerant of cold climate, but they were able to invade the Arctic during the warm climate of the Eocene.

Alternatively, Seymour et al. (2004) suggested that modern crocodiles are secondarily ectothermic and that ancestral forms were endothermic. If this is correct, endothermy would be a fundamental character of the Archosauria, which seems unlikely. Pterosaur tracks from the ancient high latitude of what is now Denali National Park could be consistent with the Archosauria being primitively endothermic, but the ability to fly, which may require endothermy in a vertebrate, is not evidence of a primitive feature. The ultimate primitive condition is undoubtedly ectothermy, with endothermy evolving multiple times (at least once and probably more than once in archosaurs, and once in mammals).

Working in such a fossil-rich setting as the Lower Cantwell Formation for a decade, and collecting a variety of different types of information, affords our group the luxury of integrating detailed sedimentological and paleobotanical data with vertebrate ichnological data (Fiorillo et al., 2015). The corroboration of these varied datasets suggests that the Cretaceous landscape was heterogenic; specifically, that some areas where pterosaur tracks were made were in more mature gallery forests with lower sedimentation rates compared to other sites, which had higher rates of sediment aggradation. The smaller pterosaur tracks are found only in areas of more mature forest. In contrast, the areas within the Lower Cantwell Formation with more open vegetation and higher-depositional rates are where the larger pterosaur tracks are found. Thus, the physical aspects of this ancient ecosystem influenced the distribution of vertebrates across the landscape, and much like modern large birds that tend to roost in open areas (e.g., Ralph, 1985) versus densely vegetated areas, during the Cretaceous larger winged vertebrates preferred more open areas, while the smaller winged vertebrates were less affected by vegetative cover and river dynamics (Fiorillo et al., 2015).

## SUMMARY

While the study of the traces of ancient organisms, or ichnology, has roots that extend back to early naturalists such as Leonardo da Vinci and Ulisse Aldrovandi, the study of fossil footprints as we have come to know it today begins in early 19th-century North America with Edward Hitchcock. The science of animal tracks and traces today, both modern and ancient, has provided tools and substance to ecological interpretations.

The most robust documented dinosaur footprint record from the high latitudes is from the Cretaceous of Alaska, and while records of dinosaur footprints are found around the state, the most thoroughly documented record comes from within national parks, including Aniakchak National Monument, Denali National Park, Wrangell–St. Elias National Park, and Yukon–Charley Rivers National Preserve. Together these park units stretch across a vast region and document the presence of an extensive ancient high-latitude terrestrial ecosystem during the later Cretaceous, one that was capable of supporting populations of large-bodied animals (i.e., dinosaurs).

Denali National Park records fifteen vertebrate ichnotaxa ranging from fossil fish traces to archosaurs. Within the Dinosauria, the theropods (including birds) show the greatest biodiversity, with at least nine ichnotaxa. Further, the fossil bird record from the Lower Cantwell Formation records two unique ichnotaxa, *Magnoavipes denaliensis* and *Gruipeda vegrundianus*. The Lower Cantwell Formation also shows evidence of diversity in the pterosaur track record by preserving tracks of two very different sizes.

The integration of sedimentological and paleobotanical datasets with the vertebrate ichnological record found in the Lower Cantwell Formation in Denali National Park shows that the landscape was heterogenic—that some areas had more mature gallery forests with lower sedimentation rates compared to other sites with higher rates of sediment aggradation with more open vegetation and higher-depositional rates. Not surprisingly, the varied physical landscape influenced the habitat preferences of at least some these animals—specifically, the smaller pterosaur tracks are found only in areas of more mature forest, while the larger pterosaurs preferred more open areas.

## REFERENCES

Alexander, R. McN. 1976. Estimates of speeds of dinosaurs. *Nature* 261:129–130.

Altendorfer, A., N. Moore, H. Komsuoglu, M. Buehler, H.B. Brown, Jr., D. McMordie, U. Saranli, R. Full, and D.E. Koditschek. 2002. RHex: A biologically inspired hexapod runner. *Journal of Autonomous Robots* 11:207–213.

Baird, D. 1957. Triassic reptile footprint faunules from Milford, New Jersey. *Bulletin of the Museum of Comparative Zoology, Harvard University* 117:449–520.

Baucon, A. 2008. Italy, the cradle of ichnology: The legacy of Aldrovandi and Leonardo. In *Ichnology in Italy*, eds. M. Avanzini and F. Petti, 15–29. Trento: Studi Trentini di Scienze Naturali–Acta Geologica, 83.

Baucon, A. 2010. Leonardo da Vinci, the founding father of ichnology. *Palaios* 25:361–367.

Bennett, M.R. and S.A. Morse. 2014. *Human Footprints: Fossilised Locomotion?* Heidelberg: Springer.

Bird, R.T. 1985. *Bones for Barnum Brown: Adventures of a Dinosaur Hunter.* Fort Worth: Texas Christian University Press.

Buatois, L.A. and M.G. Mangano. 2011. *Ichnology: Organism–Substrate Interactions in Space and Time*. New York: Cambridge University Press.

Burne, R.V. and L.S. Moore. 1987. Microbialites: Organosedimentary deposits of benthic microbial communities. *Palaios* 2:241–254.

Carvalho, I.S., L. Borghi, and G. Leonardi. 2013. Preservation of dinosaur tracks induced by microbial mats in the Sousa Basin (Lower Cretaceous), Brazil. *Cretaceous Research* 44:112–121.

Clark, J.M., T. Maryanska, and R. Barsbold. 2004. Therizinosauroidea. In *The Dinosauria*, 2nd edn, eds. D.B. Weishampel, P. Dodson, and H. Osmolska, 151–164. Berkeley: University of California Press.

Currie, P.J., G. Nadon, and M.G. Lockley. 1991. Dinosaur footprints with skin impressions from the Cretaceous of Alberta and Colorado. *Canadian Journal of Earth Sciences* 28:102–115.

Currie, P.J., D. Badamgarav, and E.B. Koppelhus. 2003. The first Late Cretaceous footprints from the Nemegt locality in the Gobi of Mongolia. *Ichnos* 10:1–13.

Dai, H., L. Xing, D. Marty, J. Zhang, W.S. Persons IV, H. Hu, and F. Wang. 2015. Microbially-induced sedimentary wrinkle structures and possible impact of microbial mats for the enhanced preservation of dinosaur tracks from the Lower Cretaceous Jiaguan Formation near Qijiang (Chongqing, China). *Cretaceous Research* 53:98–109.

Demicco, R.V. and L.A. Hardie. 1994. *Sedimentary Structures and Early Diagenetic Features of Shallow Marine Carbonate Deposits*. Tulsa, OK: SEPM (Society for Sedimentary Geology), Atlas Series, 1.

Detterman, R.L., J.E. Case, J.W. Miller, F.H. Wilson, and M.E. Yount. 1996. Stratigraphic framework of the Alaska Peninsula. *United States Geological Survey Bulletin* 1969-A:1–74.

Dilkes, D.W. 2001. An ontogenetic perspective on locomotion in the Late Cretaceous dinosaur *Maiasaura peeblesorum* (Ornithischia: Hadrosauridae). *Canadian Journal of Earth Sciences* 38:1205–1227.

Ducarme, F., G.M. Luque, and F. Courchamp. 2013. What are "charismatic species" for conservation biologists? *BioSciences Master Reviews* 1:1–8.

Dupraz, C., P.T. Visscher, L.K. Baumgartner, and R.P. Reid. 2004. Microbe-mineral interactions: early carbonate precipitation in a hypersaline lake (Eleuthera Island, Bahamas). *Sedimentology* 51:745–765.

Eiseman, C. and N. Charney. 2010. *Tracks & Sign of Insects and Other Invertebrates: A Guide to North American Species*. Mechanicsburg: Stackpole Books.

Elbroch, M. 2003. *Mammal Tracks & Sign: A Guide to North American Species*. Mechanicsburg: Stackpole Books.

Elbroch, M. and E. Marks. 2001. *Bird Tracks & Sign: A Guide to North American Species*. Mechanicsburg: Stackpole Books.

Falkingham, P.L. and S.M. Gatesy. 2014. The birth of a dinosaur footprint: Subsurface 3D motion reconstruction and discrete element simulation reveal track ontogeny. *Proceedings of the National Academy of Sciences* 111:18279–18284.

Farlow, J.O. and P. Dodson. 1974. The behavioral significance of frill and horn morphology in ceratopsian dinosaurs. *Evolution* 29:353–361.

Farlow, J.O., W. Langston, Jr., E.E. Deschner, R. Solis, W. Ward, B.L. Kirkland, S. Hovorka, T.L. Reece, and J. Whitcraft. 2006. *Texas Giants: Dinosaurs of the Heritage Museum of the Texas Hill Country*. Calgary: Dominion Exploration and Production.

Farlow, J.O. and M.G. Lockley. 1989. Roland T. Bird, dinosaur tracker: An appreciation. In *Dinosaur Tracks and Traces*, eds. D.D. Gillette and M.G. Lockley, 33–46. New York: Cambridge University Press.

Fiorillo, A.R. 2005. Turtle tracks in the Judith River Formation (Upper Cretaceous) of south-central Montana. *Palaeontologica Electronica* 8:1 MB.

Fiorillo, A.R. and T.L. Adams. 2012. A therizinosaur track from the Lower Cantwell Formation (Upper Cretaceous) of Denali National Park, Alaska. *Palaios* 27: 395–400.

Fiorillo, A.R., T.L. Adams, and Y. Kobayashi. 2012. New sedimentological, palaeobotanical, and dinosaur ichnological data on the palaeoecology of an unnamed Late Cretaceous rock unit in Wrangell–St. Elias National Park and Preserve, Alaska, USA. *Cretaceous Research* 37:291–299.

Fiorillo, A.R., M. Contessi, Y. Kobayashi, and P.J. McCarthy. 2014a. Theropod tracks from the Lower Cantwell Formation (Upper Cretaceous) of Denali National Park, Alaska, USA with comments on theropod diversity in an ancient, high-latitude terrestrial ecosystem. In *Tracking Dinosaurs and Other Tetrapods in North America*, eds. M. Lockley and S.G. Lucas, 429–439. Albuquerque: New Mexico Museum of Natural History and Science.

Fiorillo, A.R., P.L. Decker, D.L. LePain, M. Wartes, and P.J. McCarthy. 2010. A probable Neoceratopsian Manus Track from the Nanushuk Formation (Albian, Northern Alaska). *Journal of Iberian Geology* 36:165–174.

Fiorillo, A.R., F. Fanti, C. Hults, and S.T. Hasiotis. 2014b. New ichnological, paleobotanical and detrital zircon data from an unnamed rock unit in Yukon–Charley Rivers National Preserve (Cretaceous: Alaska): Stratigraphic implications for the region. *Palaios* 29:16–26.

Fiorillo, A.R., S.T. Hasiotis, and Y. Kobayashi. 2014c. Herd structure in Late Cretaceous polar dinosaurs: A remarkable new dinosaur tracksite, Denali National Park, Alaska, USA. *Geology* 42:719–722.

Fiorillo, A.R., S.T. Hasiotis, Y. Kobayashi, B.H. Breithaupt, and P.J. McCarthy. 2011. Bird tracks for the Upper Cretaceous Cantwell Formation of Denali National Park, Alaska, USA: A new perspective on ancient polar vertebrate biodiversity. *Journal of Systematic Palaeontology* 9:33–49.

Fiorillo, A.R., S.T. Hasiotis, Y. Kobayashi, and C.S. Tomsich. 2009. A pterosaur manus track from Denali National Park, Alaska Range, Alaska, USA. *Palaios* 24:466–472.

Fiorillo, A.R., Y. Kobayashi, P.J. McCarthy, T.C. Wright, and C.S. Tomsich. 2015. Reports of pterosaur tracks from the Lower Cantwell Formation (Campanian-Maastrichtian) of Denali National Park, Alaska, USA, with comments about landscape heterogeneity and habitat preference. *Historical Biology* 27:672–683.

Fiorillo, A.R., R. Kucinski, and T.R. Hamon. 2004. New frontiers, old fossils: Recent dinosaur discoveries in Alaska's National Parks. *Alaska Park Science* 3:4–9.

Fiorillo, A.R. and J.T. Parrish. 2004. The first record of a Cretaceous dinosaur from western Alaska. *Cretaceous Research* 25:453–458.

Fiorillo, A.R. and R.S. Tykoski. 2016. Small hadrosaur manus and pes tracks from the lower Cantwell Formation (Upper Cretaceous), Denali National Park, Alaska: Implications for locomotion in juvenile hadrosaurs. *Palaios* 31:479–482.

Flaig, P.P., S.T. Hasiotis, and A.R. Fiorillo. 2017. A paleopolar dinosaur track site in the Cretaceous (Maastrichtian) Prince Creek Formation of Arctic Alaska: Track characteristics and probable track-makers. *Ichnos*. doi:10.1080/10420940.2017.1337011.

Forrest, L.R. 1988. *Field Guide to Tracking Animals in Snow*. Mechanicsburg: Stackpole Books.

Fowell, S.J., P. Druckenmiller, P.J. McCarthy, R.B. Blodgett, and K. May. 2011. Paleoecology of Alaska's Jurassic Park. *Geological Society of America Abstracts with Programs* 43(5):264.

Fowler, D.W., E.A. Freedman, J.B. Scannella, and R.E. Kambic. 2011. The predatory ecology of *Deinonychus* and the origin of flapping in birds. *PLoS ONE* 6(12):e28964. https://doi.org/10.1371/journal.pone.0028964.

Galton, P.M. 1970. The posture of hadrosaurian dinosaurs. *Journal of Paleontology* 44:464–473.

Gangloff, R.A. 1998. Newly discovered dinosaur trackways from the Cretaceous Chandler Formation, National Petroleum Reserve–Alaska. *Journal of Vertebrate Paleontology* 18(Suppl 3):45A.

Gatesy, S.M. 2003. Direct and indirect track features: What sediment did a dinosaur touch? *Ichnos* 10:91–98.

Gatesy, S.M., N.H. Shubin, and F.A. Jenkins, Jr. 2005. Anaglyph stereo imaging of dinosaur track morphology and microtopography. *Palaeontologia Electronica* 8:1 MB.

Halfpenny, J. 1986. *A Field Guide to Mammal Tracking in North America.* Boulder: Johnson Printing Company.

Henderson, D. 2003. Footprints, trackways, and hip heights of bipedal dinosaurs—Testing hip height predictions with computer models. *Ichnos* 10:99–114.

Hitchcock, C.H. 1895. Edward Hitchcock. *American Geologist* 16: 133–149.

Hitchcock, E. 1858. *A Report on the Sandstone of the Connecticut Valley.* Boston: William White.

Hubbard, B.R. 1943. *Mush, You Malamutes!* New York: The America Press.

Irving, L. 1960. Birds of Anaktuvuk Pass, Kobuk, and old crow: A study in arctic adaptation. *United States National Museum Bulletin* 217:1–409.

Irving, L. 1972. *Arctic Life of Birds and Mammals, Including Man.* Berlin: Springer-Verlag.

Kendall, C.G. St. C. and P.A. d'E. Skipwith. 1968. Recent algal mats of a Persian Gulf lagoon. *Journal of Sedimentary Petrology* 38:1040–1058.

Kvale, E.P., A.D. Johnson, D.L. Mickelson, K. Keller, L.C. Furer, and A.W. Archer. 2001. Middle Jurassic (Bajocian and Bathonian) dinosaur megatracksite, Bighorn Basin, Wyoming, USA. *Palaios* 16:233–254.

Lehman, T.M. 1989. *Chasmosaurus mariscalensis*, n. sp., a new ceratopsian dinosaur from Texas. *Journal of Vertebrate Paleontology* 9:137–162.

Lesley, J.P. 1877. *Memoir of Edward Hitchcock (1793–1864).* National Academy of Sciences Bibliographic Memoirs 113–134.

Liebenberg, L. 1990. *Art of Tracking: The Origin of Science.* Cape Town: David Philip Publishers Ltd.

Lockley, M. 1991. *Tracking Dinosaurs: A New Look at an Ancient World.* Cambridge: Cambridge University Press.

Lockley, M.G. 2009. New perspective on morphological variation in tridactyl footprints: Clues to widespread convergence in developmental dynamics. *Geological Quarterly* 53:415–432.

Lockley, M.G. and A.P. Hunt. 1995. Ceratopsid tracks and associated ichnofauna from the Laramie Formation (Upper Cretaceous: Maastrichtian) of Colorado. *Journal of Vertebrate Paleontology* 15:592–614.

Lockley, M.G., T.J. Logue, J.J. Moratalla, A.P. Hunt, R.J. Schultz, and J.W. Robinson. 1995. The fossil trackway Pteraichnus is pterosaurian, not crocodilian: Implications for the global distribution of pterosaur tracks. *Ichnos* 4:7–20.

Lockley, M.G., G. Nadon, and P.J. Currie. 2004. A diverse dinosaur-bird footprint assemblage from the Lance Formation, Upper Cretaceous, eastern Wyoming: Implications for Ichnotaxonomy. *Ichnos* 11:229–249.

Lull, R.C. 1933. A revision of the Ceratopsia or Horned Dinosaurs. *Memoirs of the Peabody Museum of Natural History* 3:1–175.

Manning, P.L. 2004. A new approach to the analysis and interpretation of tracks: Examples from the Dinosauria. In *The Application of Ichnology to Palaeoenvironmental and Stratigraphic Analysis*, ed. D. McIlroy, 93–123. London: Geological Society Special Publications, 228.

Manning, P.L., D. Payne, J. Pennicott, P.M. Barrett, and R.A. Ennos. 2006. Dinosaur killer claws or climbing crampons? *Biology Letters* 2:110–112.

Manning, P.L., L. Margetts, M.R. Johnson, P.J. Withers, W.I. Sellers, P.L. Falkingham, P.M. Mummery, P.M. Barrett, and D.R. Raymont. 2009. Biomechanics of dromaeosaurid dinosaur claws: Application of x-ray microtomography, nanoindentation, and finite element analysis. *The Anatomical Record* 292:1397–1405.

Markwick, P.J. 1998. Fossil crocodilians as indicators of Late Cretaceous and Cenozoic climates: Implications for using palaeontological data in reconstructing palaeoclimate. *Palaeogeography, Palaeoclimatology, Palaeoecology* 137:205–271.

Marty, D., A. Strasser, and C.A. Meyer. 2009. Formation and taphonomy of human footprints in microbial mats of present-day tidal-flat environments: Implications for the study of fossil footprints. *Ichnos* 16:127–142.

Mayor, A. and W.A.S. Sarjeant. 2001. The folklore of footprints in stone: From Classical Antiquity to the Present. *Ichnos* 8:143–163.

Mazin, J.-M., J.P. Billo-Bruyat, P. Hantzpergue, and G. Lafaurie. 2003. Ichnological evidence for quadrupedal locomotion in pterodactyloid pterosaurs: Trackways from the Late Jurassic of Crayssac (southwestern France). In *Evolution and Palaeobiology of Pterosaurs*, eds. E. Buffetaut and J.-M. Mazin, 283–296. London: Geological Society Special Publications, 217.

McCrea, R.T., M.G. Lockley, and C.A. Meyer. 2001. Global distribution of purported Ankylosaur track occurrences. In *The Armored Dinosaurs*, ed. K. Carpenter, 413–454, Bloomington: Indiana University Press.

Milàn, J. and R.G. Bromley. 2006. True tracks, undertracks and eroded tracks, experimental work with tetrapod tracks in laboratory and field. *Palaeogeography, Palaeoclimatology, Palaeoecology* 231:253–264.

Milàn, J. and R.G. Bromley 2008. The impact of sediment consistency on track and undertrack morphology: Experiments with emu tracks in layered cement. *Ichnos* 15:18–24.

Murie, O.J. 1954. *A Field Guide to Animal Tracks*. Boston: Houghton Mifflin Company.

Nadon, G.C. 2001. The impact of sedimentology on vertebrate track studies. In *Mesozoic Vertebrate Life*, eds. P.J. Currie, D.H., Tanke, D.H., K. Carpenter, and M.W. Skrepnick, 395–407. Bloomington: Indiana University Press.

Noffke, N. 2010. *Geobiology: Microbial Mats in Sandy Deposits from the Archean Era to Today*. Heidelberg: Springer.

Noffke, N., G. Gerdes, T. Klenke, and W.E. Krumbein. 2001. Microbially induced sedimentary structures—A new category within the classification of primary sedimentary structures. *Journal of Sedimentary Research* 71:649–656.

Osborn, H.F. 1912. Integument of the iguanodont dinosaur *Trachodon. American Museum of Natural History Memoirs* 1:33–54.

Pearce, N.J.G., J.A. Westgate, S.J. Preece, W.J. Eastwood, and W.T. Perkins. 2004. Identification of Aniakchak (Alaska) tephra in Greenland ice core challenges the 1645 BC date for Minoan eruption of Santorini. *Geochemistry, Geophysics, Geosystems* 5:Q03005, doi:10.1029/2003GC000672.

Peck, R.M. 2012. *A Glorious Enterprise: The Academy of Natural Sciences of Philadelphia and the Making of American Science*. Philadelphia: University of Pennsylvania Press.

Pemberton, S.G., M.K. Gingras, and J.A. MacEachern. 2007. Edward Hitchcock and Roland Bird: Two early Titans of vertebrate ichnology in North America. In *Trace Fossils: Concepts, Problems, Prospects*, ed. W. Miller III, 32–51. Amsterdam: Elsevier.

Porada, H., E. Bouougri, and J. Ghergut. 2007. Hydraulic conditions and mat related structures in tidal flats and coastal sabkhas. In *Atlas of Microbial Mat Features Preserved within the Clastic Rock Record*, eds. J. Schieber, J., B.K. Bose, P.G. Eriksson, S. Banerjee, W. Altermann, and O. Catuneau, 258–265. Amsterdam: Elsevier.

Ralph, C.J. 1985. Habitat association patterns of forest and steppe birds in northern Patagonia, Argentina. *Condor* 87:471–483.

Razzolini, N.L., B. Vila, D. Castanera, P.L. Falkingham, J.L. Barco, J.I. Canudo, P.L. Manning, and A. Galobart. 2014. Intra-trackway morphological variations due to substrate consistency: The El Frontal dinosaur tracksite (Lower Cretaceous, Spain). *PLoS ONE* 9(4): e93708. doi:10.1371/journal.pone.0093708.

Rezendes, P. 1999. *Tracking & the Art of Seeing: How to Read Animal Tracks and Sign*, 2nd edn. New York: Quill, HarperCollins Publishers.

Salisbury, S.W., A. Romilio, M.C. Herne, R.T. Tucker, and J.P. Nair. 2016. The Dinosaurian Ichnofauna of the Lower Cretaceous (Valanginian–Barremian) Broome Sandstone of the Walmadany Area (James Price Point), Dampier Peninsula, Western Australia. *Journal of Vertebrate Paleontology* 36(suppl 1):1–152.

Sarjeant, W.A.S. (ed.) 1983. *Terrestrial Trace Fossils*. Stroudsburg, Hutchinson Ross Publishing Company.

Seilacher, A., L.A. Buatois, and M.G. Mángano. 2005. Trace fossils in the Ediacaran–Cambrian transition: Behavioral diversification, ecological turnover and environmental shift. *Palaeogeography, Palaeoclimatology, Palaeoecology* 227:323–356.

Senter, P. 2012. Forearm orientation in Hadrosauridae (Dinosauria: Ornithopoda) and implications for museum mounts. *Palaeontologia Electronica* 15, 10pp, palaeoelectronica. org/content/2012-issue-3-articles/324-hadrosaurid-forearm

Sergio, F., I. Newton, L. Marchesi, and P. Pedrini. 2006. Ecologically justified charisma: Preservation of top predators delivers biodiversity conservation. *Journal of Applied Ecology* 43:1049–1055.

Seymour, R.S., C.L. Bennett-Stamper, S.D. Johnston, D.R. Carrier, and G.C. Grigg. 2004. Evidence for endothermic ancestors of crocodiles at the stem of Archosaur evolution. *Physiological and Biochemical Zoology* 77:1051–1067.

Skalski, J. 1991. Using sign counts to quantify animal abundance. *Journal of Wildlife Management* 55:705–715.

Smith, J.B. and J.O. Farlow. 2003. Osteometric approaches to trackmaker assignment for Newark Supergroup ichnogenera Grallator, Anchisauripus, and Eubrontes. In *The Great Rift Valleys of Pangea in Eastern North America, Vol. 2: Sedimentology, Stratigraphy, and Paleontology*, eds. P.M. LeTourneau and P.E. Olsen, 273–292. New York: Columbia University Press.

Steinbock, R.T. 1989. Ichnology of the Connecticut Valley: A vignette of American science in the early 19th century. In *Dinosaur Tracks and Traces*, eds. D.D. Gillette and M.G. Lockley, 27–32. New York: Cambridge University Press.

Tomsich, C.S., P.J. McCarthy, A.R. Fiorillo, D.B. Stone, J.A. Benowitz, and P.B. O'Sullivan. 2014. New zircon U-Pb ages for the lower Cantwell Formation: Implications for the Late Cretaceous paleoecology and paleoenvironment of the lower Cantwell Formation near Sable Mountain, Denali National Park and Preserve, central Alaska Range, USA. In *Proceedings of the International Conference on Arctic Margins VI, Fairbanks, Alaska, May 2011*, eds. D.B. Stone, G.K. Grikurov, J.G. Clough, G.N. Oakey, and D.K. Thurston, 19–60. St. Petersburg: VSEGEI.

Van Valkenburgh, B. and R.E. Molnar. 2002. Dinosaurian and mammalian predators compared. *Paleobiology* 28:527–543.

Witton M.P. and D. Naish. 2008. A Reappraisal of Azhdarchid pterosaur functional morphology and paleoecology. *PLoS ONE* 3(5):e2271. doi:10.1371/journal.pone.0002271.

Zanno, L.E. 2010. A taxonomic and phylogenetic re-evaluation of Therizinosauria (Dinosauria: Maniraptora). *Journal of Systematic Palaeontology* 8:503–543.

# 6 Overview of Cretaceous Plants from Alaska

## INTRODUCTION

Plants fall into two groups of species: those without seeds and those with seeds. The former group is very ancient, predating the Cretaceous, appearing 450 million years ago as plants first began colonizing land during the Devonian Period. Seedless plants include liverworts, mosses, horsetails, and ferns. Seed plants are also ancient, first appearing 300 million years ago, and include cycads, ginkgos, and conifers. Together these form a group known as gymnosperms (Figure 6.1). This second group of seed-bearing plants also includes the flowering plants, or angiosperms.

Conifers constitute the biggest group of gymnosperms, and modern conifer forests are dominant over large areas of landscape, particularly in cooler climates, as in much of modern Alaska. And like today, conifers were prevalent over much of the landscape in earlier geologic time. There is some debate regarding the origin of flowering plants and what constitutes the angiosperms. What is not questioned, however, is that angiosperms become a prominent part of the global flora beginning in the Early Cretaceous, 120 million years ago. These earlier angiosperms were likely weedy, and based on the fossil wood record, the changeover in floral dominance occurred later such that angiosperms did not become forest dominant until toward the end of the Cretaceous (Peralta-Medina and Falcon-Lang, 2012).

The Cretaceous rocks of Alaska contain an abundance of extraordinary fossil floras, ages from 113 to 66 million years ago, and the overwhelming majority of these floras are dominated by fossil plants with seeds. These Alaskan fossil floras have been known since the 1950s when R.L. Langenheim, Jr. and C.J. Smiley followed up on initial reports of Cretaceous amber (fossilized tree sap), and began a study of Cretaceous sedimentary units in northern Alaska along the Kaolak River (Langenheim et al., 1960; Smiley, 1966). While the early efforts focused on understanding the context of the amber, it became clear that additional work on the fossil plant remains was also insightful (Smiley, 1966). Since these early studies, the Cretaceous plants of Alaska have been studied extensively and many of these floras correlate with those in northeastern Russia, providing opportunities for broad regional syntheses (e.g., Smiley, 1966, 1969a,b, 1972a,b; Krassilov, 1975; Spicer et al., 1987; Spicer and Herman, 1996, 2001, 2010; Herman, 2002, 2004, 2007a,b,c; Spicer, 2003; Tomsich et al., 2010; Herman et al., 2016). Though work in Cretaceous rocks of northeastern Russia was initiated in 1912–1913, systematic mapping and study of fossil plants from these Russian Cretaceous rock units began in the late 1940s and in earnest in the 1950s (Herman, 2013). Together the deposits of

1 cm

**FIGURE 6.1** An example of the gymnosperm *Metasequoia* from the Lower Cantwell Formation of Denali National Park, Alaska. (Specimen DENA 34127.)

northeastern Russia and northern Alaska have been termed the North Pacific Region (e.g., Herman, 2007a,b,c, 2013; Herman et al., 2016).

Smiley's original work eventually amassed a collection of some 10,000 specimens from northern Alaska (Herman, 2004), which has only grown through the efforts of various other workers. While most of the paleobotanical record is known from megafloral remains, some rock units also contain rich assemblages of fossil pollen (Fredriksen, 1990; Fiorillo et al., 2010; Flaig et al., 2013).

The paleofloras are from an ancient ecosystem with no modern analog. They existed under a polar light regime similar to that of the present day but experienced temperatures far warmer than current temperatures experienced at these latitudes. Broadly speaking, temperate forests thrived during the Cretaceous in the ancient Arctic at latitudes that may have been as high as 85°N (Witte et al., 1987; Besse and Courtillot, 1991). However, rather than being a forest realm where plants were physiologically homogenous, there was likely significant variation in growing season across these high latitudes, as trees growing at 65°N today have a growing season that is a few months longer than the one experienced by latitude 80°N in the Cretaceous (Beerling and Osborne, 2002). So, while the temperatures in which Cretaceous ecosystems existed were warmer than what is experienced in the region today due to increased concentrations of $CO_2$ during the Mesozoic, the Arctic light regime would have been similar to the modern regime because it is controlled by the Earth's tilt and yearly orbit around the sun (Chapter 1).

Therein lies the conundrum. What adaptations allowed extensive angiosperm forests to grow at high latitudes in the Cretaceous under the extraordinary light conditions of long periods of light and long periods of dark? Today, one of the primary challenges for seed-producing plants living in this highly seasonal light regime is to complete seed generation within the short Arctic summer, and to accommodate seed production there is the rapid initiation of growth in the spring, which in modern high-latitude seed-bearing plants begins just weeks before summer solstice (Savile, 1972). While within the moisture-limiting seasonal environment of lower latitude deserts, some plants have adapted by developing subterranean storage organs that provide immediate resumption of growth when water becomes available, such organs are not commonly found in the Arctic because soils often remain very cold well into the summer months (Savile, 1972). For these Arctic plants, plant growth requires immediate photosynthesis, and presumably these same constraints were in operation during the Cretaceous.

The angiosperm invasion of the Arctic began in the later part of the Albian (approximately 113 million years ago) with the flowering plants first occupying disturbed riparian areas, then later expanding their role into the understory of coniferous woodlands (Retallack and Dilcher, 1986; Herman, 2002, 2013). Some argued that trees within ancient polar ecosystems were deciduous as many angiosperms are today to avoid losing carbon through respiration during the long winter (see Osborne et al., 2004 for review).

Today, evergreen conifers dominate the boreal zone. In North America, the most abundant conifers are the evergreen spruces (*Picea*), pines (*Pinus*), and firs (*Abies*), but in Asia the dominant conifers are deciduous larches (*Larix*), which cover much of the interior of Siberia. Evergreens replace their needles every few years as opposed to the yearly loss of foliage in deciduous trees, reducing their nutritional needs and while their photosynthetic efficiency is less than that of deciduous woody plants (Pielou, 1988; Schultz, 2005). Deciduousness is one strategy for coping with strong seasonality, but a deciduous forest may be up to twice as expensive in terms of carbon loss to the plant through the shedding of leaves, compared to wintertime respiration in an evergreen forest (Royer et al., 2003; Osborne et al., 2004; Brentnall et al., 2005).

Like the modern boreal zone, the vegetation of the Cretaceous of Antarctica, the antipodal analog of the Arctic, was vegetated and dominated by evergreens (Cantrill and Poole, 2012). In contrast, the Cretaceous forests of ancient Alaska are considered to have been dominated by deciduous conifers along with much less abundant angiosperms (Spicer, 2003; Herman et al., 2016). So, the majority of woody plants living in the Cretaceous Arctic seem to have solved the ecophysiological problem posed by the long winters and concomitant periods of reduced light by being deciduous. The asymmetrical response at the poles to similar climatic conditions, and specifically temperature and light regimes, is intriguing. This asymmetry reflects the paleobiogeography of each polar flora. The ancient Arctic was centered on an ocean basin, while the ancient Antarctic was centered on a large landmass at the other end of the globe. These geographic factors also played a role in driving the dominant ecophysiology within these respective polar floras.

Common components other than deciduous trees in the Cretaceous Alaskan flora were herbaceous shrubs and ferns. Modern annual herbaceous plants die as winter approaches and then overwinter as seeds, while perennial herbaceous plants die back to rhizomes, or root masses, then sprout again with light and warmer weather. So, the Cretaceous Arctic flora seems to have survived the long winter months by a combination of deciduousness in trees and die back in annual plants and herbaceous perennials. This raises a paradox: What did the diverse and abundant dinosaurs—more importantly the herbivorous dinosaurs—do for food during the winter months?

While various lines of vertebrate fossil evidence suggest that these latest Cretaceous dinosaurs were year-round residents of the ancient Arctic, such as biomechanical analysis of the limb proportions of the Arctic hadrosaur remains of *Edmontosaurus* that suggests the juveniles recovered from northern Alaska were likely too small to migrate ahead of the descending winter darkness (Chapter 4; Fiorillo and Gangloff, 2001; Bell and Snively, 2008), the fossil plant record provides some very important clues to the paleobiology of the terrestrial vertebrates of the ancient Arctic that can seem incongruent with the presence of such a rich record of these animals (Herman et al., 2016; Spicer et al., 2016). In their quantitative reconstruction of temperatures and qualitative reconstruction of precipitation for the Cretaceous Arctic, and specifically the Alaskan Arctic, Spicer and others (2016) suggest that dinosaurs nested under cool, damp conditions and that growth must have been rapid during the summer months when vegetable food and calories were plentiful, allowing the herbivores to stock up for the winter when food resources were limited.

## THE NANUSHUK FORMATION

In northern Alaska, one of the oldest terrestrial dinosaur-bearing rock units is the Albian (112-million-year-old) to Cenomanian (95-million-year-old) aged Nanushuk Formation, which contains both fossil plants and fossil vertebrates. It also contains volcanic ash, which has provided improved geochronology through isotopic dating (Shimer et al., 2016). Shimer and others (2016) published radiometric dates of $102.6 \pm 1.5$ million years ago (late Albian) and $98.2 \pm 0.8$ million years ago (early Cenomanian) for the Nanushuk Formation from bentonites recovered in the central part of the North Slope of Alaska near Umiat Mountain. The volcanic source of these tuffaceous sediments was the Okhotsk-Chukotka volcanogenic belt in northeastern Asia (Herman, 2004). The Nanushuk flora is composed of ferns, ginkgos and their relatives, cycads and their relatives, conifers, and angiosperms (Herman, 2004). The rise of angiosperms is considered to have been rapid, with platanoids, the relatives of modern sycamore trees, becoming abundant as the angiosperms take their place (Smiley, 1966). The Nanushuk localities were near sea level, along the border of the cool polar ocean (Smiley, 1969a).

The thickest exposures of non-marine sections of the Nanushuk Formation are in the western part of northern Alaska. Spicer and Herman (2001) studied fossil plants of the Nanushuk Formation within this western region along a significant stretch of the north-flowing 260-kilometer-long river Kukpowruk River (Figure 6.2). They recognized some 57 plant taxa, identified largely from leaves but also including some taxa known from reproductive structures (fruits and spore-bearing structures) of unknown taxonomic affinity (for example, *Kenella filatovii, Stenorachis striolatus*). The flora they reported was diverse and included possible liverworts, Equisetales, ferns, cycadophytes, ginkgoaleans, possible czekanowskialeans, conifers, gymnosperms incertae sedis, and angiosperms, with the ferns and conifers exhibiting the most diversity (11 and 12 species, respectively). The floras they discovered showed strong taxonomic similarities to floras of comparable age in northeastern Russia as well as to the east in the central and eastern part of the North Slope of Alaska (Spicer and Herman, 2001; Herman, 2007a,b,c).

Consistent associations between sedimentary facies and specific taxa were noted, which were then interpreted as representing distinct plant communities. One community was a marginal marine and early successional marsh or heath dominated by the horsetail *Equisetites*, considered to be the initial colonizer, and *Birisia* ferns. A later-stage successional stage is characterized by the fern *Arctopteris* and several *Nilssonia*

**FIGURE 6.2** View looking north of exposures of the Nanushuk Formation along the Kukpowruk River in northwestern Alaska.

species. *Nilssonia* was a plant with fronds that looked generally like a cycad but was from an unrelated and extinct group of plants, the Bennettitales, or the Nilssoniales (see below). Spicer and Herman (2001) follow earlier interpretations of *Nilssonia* as a shrub-like plant with slender stems bearing whorls of leaves on short or dwarf shoots (e.g., Spicer and Herman, 1996). Trees within this floral community were uncommon, but the *Equisetites*/*Birisa* association likely continued within the understory during the later successional stages of community development after trees moved in.

A second floral community described by Spicer and Herman (2001) is a shrubby to forested river margin community dominated by the non-flowering tree, the *Ginkgo*. This riparian community, dominated by these trees, occurred on coarse-grained, well-drained coastal plain substrates.

In contrast to this well-drained environment, mire forest communities were dominated by the gymnosperms *Pityophyllum* and *Podozamites*. *Pityophyllum* is a conifer but it is unclear whether it grew as a tree or a shrub. *Podozamites*, a likely cycad, is dominant in areas with poor coal development while *Pityophyllum* is dominant in areas with more coal, indicating differences in paleohydrology and suggesting *Pityophyllum* preferred moister environments. (Spicer and Herman, 2001).

Spicer and Herman (2001) recovered angiosperms from only one site during their 1998 field season, although Smiley (1969a) recorded angiosperms from an additional area along the Kukpowruk River. Despite their efforts in their 1996 field season, Spicer and Herman (2001) were unable to recover angiosperms at Smiley's site. Thus, in their reconstruction of the Nanushuk floral communities along this river, Spicer and Herman (2001) recognized a co-occurrence of early angiosperms with conifers as a community that was present but decidedly rare and local in extent. Recent work (2015–2017) by field parties from the Perot Museum of Nature and Science and the University of

Alaska Department of Geology reveals that the trees within these woodlands were widely spaced, separated by 10s of meters, suggesting open, low-density woodlands, in contrast to the tree spacing observed in the Chignik Formation (see below).

Northeast of the Kukpowruk River is the Kaolak River. The name Kukpowruk is Inupiat for "fairly large river." The Kaolak River is a much smaller river. It drains a gently rolling landscape with vegetated bluffs that reach some 15–20 meters high. The well-entrenched drainage combined with the tundra blanketing the landscape produces limited exposures of the Nanushuk Formation. The Kaolak drainage contains a significant array of floral and faunal remains comparable in age to those along the Kukpowruk (e.g., Fiorillo, 2014; Herman et al., 2016). Herman et al. (2016) referred the Kaolak flora as Turonian in age, but ash dates from a Perot Museum of Nature and Science expedition in 2004 yielded a $^{40}Ar/^{39}Ar$ age of approximately 95 million years ago, placing this part of the section within the Cenomanian.

From megafloral remains, Herman et al. (2016) recognized a decrease in fossil plant diversity but acknowledge that this may be the result of sampling bias. For example, they report only two taxa of ferns from the Kaolak Flora and no ginkoaleans or cycadophytes. Supporting their suggestion that sampling may have a role in the decrease in biodiversity, pollen in sediment samples taken from exposures along the Kaolak River produced over three dozen different spore and pollen types (N = 37), including several ferns (including tree ferns). Fern spores comprise nearly 30% of the assemblage (Table 6.1). Further, the pollen corroborates a minimum age of Cenomanian (approximately 95 million years ago) for the sections exposed along the Kaolak River (Zippi, personal communication, 2006).

## TABLE 6.1
## Major Components of Palynoflora per Probable Botanical Affinities

| Botanical Affinity | Percent of Assemblage |
| --- | --- |
| Pinaceae—conifer | 44.3 |
| Cyatheaceae—tree ferns | 18.6 |
| Lycopodiaceae—clubmosses | 8.6 |
| Selaginellaceae—spikemosses | 6.4 |
| Osmundaceae—ferns | 5.0 |
| Polypodiaceae—ferns | 5.0 |
| Indet. spore | 4.3 |
| Schizaeaceae—ferns | 4.3 |
| Indet. conifer | 2.1 |
| Sphagnaceae—moss | 1.4 |

*Note:* The botanical terms reference taxonomic families and common names. The groupings are listed by relative abundance. The sample is from the Nanushuk Formation and was taken along the Kaolak River during a Perot Museum of Nature and Science expedition in 2004.

A correlative flora for the Nanushuk Formation in northern Russia shows much higher diversity (that is, 50 taxa of ferns, ginkgoaleans, conifers, and angiosperms; Herman et al., 2016). In fact, despite the seemingly lower diversity within parts of the Nanushuk Formation, the floras taken collectively around the northern polar region indicate that Arctic forests were at their height in taxonomic diversity from the Cenomanian–Coniacian, window of time from about 100 million years ago to approximately 86 million years ago (Herman et al., 2016).

## ON THE ISSUE OF *NILSSONIA*

*Nilssonia* is a fossil plant found in the Cretaceous Alaskan high latitudes and is thought to be evolutionarily related to modern cycads, but cycads today are found only in the tropics and subtropics. The fact that a cycad-like plant (Figure 6.3) is known from the Cretaceous of the ancient high latitudes clearly has implications for its paleoclimatological setting, but the issues related to the taxonomy of this genus of plant require some elaboration.

*Nilssonia* is a fossil plant with not only an intellectually challenging legacy of understanding its paleobiology but also in understanding the spelling of the genus. Peppe et al. (2007) discussed the history of the genus name. In referring to the International Code of Botanical Nomenclature, they point out that in Chapter VII, Section 1, Article 60.1 (St. Louis Code, Greuter et al., 2000, p. 92), "the original

**FIGURE 6.3** Cretaceous cycad-like plant fossil from the Lower Cantwell Formation of Denali National Park, Alaska. (Photograph courtesy of T. Colby Wright, University of Alaska.)

name or epithet is to be retained, except for the correction of typographical or ortho-
graphical errors."

Brongniart (1825) named the genus *Nilsonia* after the Swedish naturalist Sven
Nilsson, though the second "s" was omitted from Nilsson's name. In 1841, Hisinger
changed the spelling to *Nilssonia*, and that spelling has subsequently been followed
through as the spelling used by modern workers (Harris, 1961; Spicer and Herman, 1996,
2010; Pott et al., 2012). The International Code of Botanical Nomenclature notwithstand-
ing, given the overwhelming use of the genus name with a double "s" combined with the
fact that this is not the place to meaningfully contribute to the discussion, for the following I have chosen to use the popular modified spelling, *Nilssonia*, for the genus.

Paleobiologically, *Nilssonia* is based upon leaves having a suite of characters known
to be present in living or true cycads (Order Cycadales) rather than cycadeoids—which
includes extinct and different orders of plants (the Bennettitales or the Nilssoniales), but
shares leaf morphology that is grossly similar to the leaves of cycads. The guard cells
surrounding the stomata, or respiratory pores, on the leaf epidermis are derived from
a common cell, and the subsidiary cells that surround the guard cells are derived from
other epidermal cells. This condition is found in some gymnosperms and is typical of
members of the Cycadales. Additional characteristics of *Nilssonia* include thin cuticle
in the guard cells, irregular orientation of stomata, straight-walled epidermal cells, and
epidermal cells not oriented into rows. *Nilssonia* is the only genus known to have this
suite of characteristics (Stewart and Rothwell, 1993), which makes its higher-level systematic placement difficult to resolve. Moreover, leaves, stems, seeds, pollen, and other
parts of the same plant may not be found together or may be mixed with other species in
the fossil record, and the association of these parts is often not straightforward—challenges that influence the taxonomic debate (e.g., Crane et al., 2004). Such a challenge
has resulted in the genus *Nilssonia* being placed within the Order Bennettiales, the
Order Nilssoniales, and even the Order Cycadales, the modern cycads. Regardless of
the placement within these orders of plants, it is generally considered that *Nilssonia* is
taxonomically very close to modern cycads.

Spicer and Herman (2001) interpret *Nilssonia* as growing like a shrub, and while
some living cycads, though taxonomically unrelated, grow in shrub form, others
grow as trees. Therefore the interpretation of the growth habit of this fossil taxon
based on modern forms remains open. It is also unclear whether *Nilssonia* was
deciduous or evergreen like living cycads.

Modern cycads retain their leaves year round (Cockell et al., 2008), but they live
where it is warm. Plants in cold, dark situations, such as experienced today during winters at high latitudes, can survive through decreased metabolic rates and
cool temperatures (sensu Read and Francis, 1992) even though the darkness prevents
photosynthesis (Cockell et al., 2008). If *Nilssonia* plants were indeed evergreen,
maintaining the leaves through a long, dark, but relatively warm winter would have
caused an intolerably high respiratory drain on the plant, particularly for young
plants and seedlings (Cockell et al., 2008). The implication is that deciduous growth
is the favored condition for *Nilssonia*.

*Nilssoniocladus* is a related plant described originally from the lowermost
Cretaceous, the Oguchi Formation (approximately 145–135 million years ago) in of
Japan (Kimura and Sekido, 1975). Specimens from Alaska have persistent circular

leaf scars interpreted as short shoot abscission points, meaning that the plant was deciduous and shed its leaves (Spicer and Herman, 1996). In addition, the original specimens described by Spicer and Herman (1996) are associated with mats of *Nilssonia* leaves, leading them to conclude that the biology of these Cretaceous cycad-like plants was likely different from that of modern cycads and that these fossil forms represent a subset of the group that had a wider range of environmental tolerances that allowed them to be well adapted to unique temperature and light regimes of the ancient Arctic.

## THE PRINCE CREEK FORMATION

The Prince Creek Formation is a Late Cretaceous to Paleocene non-marine to marine sedimentary unit (Flaig et al., 2013) in northern Alaska. The original stratigraphic work on this rock unit (Gryc et al., 1951) had extended it into the Coniacian (approximately 89–86 million years ago) in the form of the Tuluvak Tongue of the Prince Creek Formation, which has a diverse flora (Herman, 2013; Herman et al., 2016). However, Mull et al. (2003) revised the stratigraphic nomenclature of the Cretaceous and Tertiary rock units of northern Alaska, abandoning the term Tuluvak Tongue and elevating the unit to formation status. Within this revised definition of the Prince Creek Formation, megafloral diversity of the Prince Creek Formation decreases from 33 to 13 taxa by the Campanian, and by the Maastrichtian woody angiosperms, ginkgophytes. and cycadophytes had all but disappeared (Herman, 2013; Herman et al., 2016).

Fossil wood was among the first plant material to be examined in detail for environmental reconstruction for the Prince Creek Formation (Spicer and Parrish, 1990). Xylem is a tissue found in vascular plants that serves the purpose of transporting water and some nutrients from the roots to other parts of the vascular plant such as shoots and leaves. Within the xylem are elongated cells called tracheids, which serve as supporting and transportation tissue. The thin-walled nature and large lumina of the tracheids from the wood specimens collected from the Prince Creek Formation compared favorably with the tracheid proportions of conifers growing without environmental stress, in contrast to thick-walled tracheids that develop in vascular plants that experience water stress (Spicer and Parrish, 1990). From this pattern in the fossil wood, Spicer and Parrish (1990) suggested that these trees from the Prince Creek Formation likely did not experience periods of prolonged freezing.

The pollen record for the Prince Creek Formation is exceedingly rich (Fiorillo et al., 2010; Flaig et al., 2013). The locally derived fossil pollen from the ceratopsian-dominated Kikak-Tegoseak Quarry showed a high-relative abundance of only a few non-arboreal species of plants (i.e., ferns) and a very low abundance of aquatic plants and freshwater algae, as well as low absolute and relative abundance of Taxodiaceae, or cypress swamp conifers (Fiorillo et al., 2010). This pollen assemblage suggests that the landscape was well drained and contained little standing water, and that conifer forests spread across the coastal plain with broad-leafed angiosperms in the riparian areas (Fiorillo et al., 2010).

In a detailed integration of paleopedology and palynology, Flaig et al. (2013) were able to reconstruct specific coastal plain local environments of the Prince Creek

Formation along a roughly 50-kilometer stretch along the Colville River. Many of these local environments are dominated by specific flora. For example, lake margin soils were dominated by algae such as *Pediastrum* and *Pterospermella* as well as *Taxodiaceaepollenites*, a lowland tree determined as such by its association with lower-delta plain soils (Flaig et al., 2013). Swamp margin environments were dominated by the algae *Botryococcus*, *Pediastrum*, *Leiospheres*, the fern *Psilatriletes*, and the angiosperm *Aquatiapollenites*. Soils found on ancient point bars of the Prince Creek Formation were dominated by *Laevigatosporites* ferns and bisaccate conifer pollen, while levee deposits were dominated by bisaccate conifer pollen and that of herbaceous shrubs and ferns (Brandlen, 2008; Fiorillo et al., 2010, 2013). Crevasse splay environments were dominated by the lowland tree *Taxodiaceaepollenites*, bisaccate conifer pollen, the fern *Laevigatosporites*, the algae *Sigmapollis*, and fungal hyphae. Last, undifferentiated lower-delta environments were dominated by *Botryococcus*, *Pediastrum*, and *Sigmapollis* algae, *Laevigatosporites* ferns, the lowland tree *Taxodiaceaepollenites*, and bisaccate conifer pollen.

## THE LOWER CANTWELL FORMATION

The Cantwell Formation was named in the very early part of the 20th century as part of a reconnaissance survey to understand the topography of the region as a means for mapping potential routes for future railway lines and to explore for mineral resources (Eldridge, 1900). The formation was named for coarse-grained sedimentary units along what was then called the Cantwell River, but is now known as the Nenana River (Eldridge, 1900). It wasn't until many years later that the fossil plant potential of the Cantwell Formation became more apparent (Moffitt and Pogue, 1915; Chaney, 1937; Wolfe and Wahrhaftig, 1970). Based on these paleobotanical studies, the age of the Cantwell Formation has been variously considered Eocene (Moffitt and Pogue, 1915), Early Cretaceous (Chaney, 1937), and Paleocene (Wolfe and Wahrhaftig, 1970). A more recent study of the fossil flora from the Lower Cantwell Formation of Denali National Park in the central Alaska Range places that part of the formation in the Lower Maastrichtian (Tomsich et al., 2010). The recent discovery of volcanic ashes in midsection of this rock unit found along the Sable Mountain in Denali has yielded zircon U–Pb ages of $71.5 \pm 0.9$ and $71.0 \pm 1.1$ million years ago, indicating the rock unit straddles the Campanian-Maastrichtian boundary (Tomsich et al., 2014). Thus, Cantwell Formation correlates well with the Prince Creek Formation of northern Alaska and allows for a regional view of the distribution of plant communities during this point in geologic time.

Unlike the early Maastrichtian Prince Creek Formation, the Cantwell Flora has a relatively diverse assemblage of plants (Tomsich et al., 2010), a biodiversity pattern that is most easily explained by its lower latitude and relative nearness to shorelines (Herman et al., 2016). Non-angiosperm megafloral remains include horsetails (Figure 6.4), ferns, deciduous conifers *Metasequoia*, *Glyptostrobus*, *Tumion*, *Cephalotaxopsis*, possibly *Parataxodium*, possibly *Pityophyllum*, and cf. *Pseudolarix* (Tomsich et al., 2010). In their detailed study of the paleoflora based on leaves and leaf fragments, Tomsich and others (2010) grouped 95 angiosperm leaf

**FIGURE 6.4** Example of a fossil horsetail from the Lower Cantwell Formation of Denali National Park, Alaska. This horsetail also has nitrogen-fixing structures (circled) attached to the stalk (arrow). (Specimen DENA 20974.)

**FIGURE 6.5** Example of an angiosperm leaf from the Lower Cantwell Formation of Denali National Park, Alaska. This leaf corresponds to morphotype C of Tomsich and others (2010), a morphotype attributed to alder-like trees. (Specimen DENA 13309.)

impressions into 19 leaf morphotypes (Figure 6.5), indicating that the angiosperms were relatively diverse as well.

In contrast to these megafloral remains, there is only a sparse palynomorph assemblage (Zippi, 2004, 2005; Tomsich et al., 2010). This part of the plant record from the Lower Cantwell Formation contains a few angiosperm and conifer pollen taxa (e.g., *Taxodiaceaepollenites*, *Piceapollenites*), horsetail spores, and a variety of fern spores (e.g., *Deltoidospora*, *Dictyophyllidites*, *Laevigatosporites*), and sphagnum and spike moss spores (e.g., *Stereisporites*, *Acanthotriletes*). There is also a scant record of hyphae (Zippi, 2004), the long filamentous branching structures of vegetative growth in fungi.

The flowering plants were dominant along channels and levees as part of a mixed forest assemblage on the ancient floodplains and lake margins (Tomsich et al., 2010). A number of different conifers probably inhabited the more distal floodplain and upland environments (Tomsich et al., 2010). Distinct floral assemblages can be found within different deposition environments in the Lower Cantwell Formation. Tomsich et al. (2010) attributed the abrupt floral changes within the rock unit to rapid sediment buildup within braided river floodplains that interfingered with distal alluvial fans, resulting in the stacking of distinct floras. The floral diversity of the Lower Cantwell Formation is similar to the younger Late Maastrichtian diversity documented in Russia (Herman et al., 2016).

## THE CHIGNIK FORMATION

The rocks that now comprise the unit named the Chignik Formation of the Alaska Peninsula were originally of interest due to the common occurrence of coal seams—specifically exposures along the Chignik River (Dall, 1896). While other studies followed, Stanton and Martin (1905) added to this growing geologic awareness of these rocks by providing lists of fossil plants and marine invertebrates, as well as correlating this unit with other Upper Cretaceous units throughout the region.

Fossil leaves are abundant in the Chignik Formation as well as carbonized fossil wood. Hollick (1930) identified 49 angiosperm leaf forms from this rock unit, and his study remains the definitive work on the megafloral remains from the Chignik Formation. More recent work on this rock unit that documented the dinosaur tracks (Fiorillo and Parrish, 2004), discussed in Chapter 5, also yielded additional unpublished megafloral remains that include the taxon *Trapa*? (Figure 6.6), an aquatic fern, though its taxonomic affinity is not clear. This plant is reported from the section of the Prince Creek Formation that corresponds to the Campanian-Maastrichtian boundary (approximately 70 million years ago; Parrish and Spicer, 1988; Spicer and Parrish, 1990), so its presence in the Chignik Formation may have biostratigraphic significance.

The palynological record, like the record from the Lower Cantwell Formation of Denali National Park, is sparse (Zippi, 2004). The Chignik Formation fossil pollen record includes an angiosperm taxon (*Triporopollenites*), a conifer taxon (*Taxodiaceaepollenites*), a variety of fern spores (e.g., *Deltoidospora, Dictyophyllidites, Osmundacidites, Cicatricosisporites*), and clubmoss and spikemoss spores (*Lycopodiumsporites, Acanthotriletes*).

On at least one stratigraphic horizon within this rock unit, fossil leaves, upright trees, and dinosaur footprints are found together. These trunks are generally less than 1 m in height and less than 20 cm diameter, but clearly have been compressed laterally (Figure 6.7). Most of the tree bases can be traced into the associated root systems that spread along the base of the bed, although extensions of the root systems are not always visible in the underlying siltstone. The spacing between these trunks is 1–6 m. Thin laminae of carbonaceous shale laterally between some of the upright trees contain leaves, and many of the angiosperm leaves exhibit some evidence of herbivorous insect damage—specifically, leaf miners.

**FIGURE 6.6** An example of *Trapa*? leaves from the Chignik Formation of Aniakchak National Monument, Alaska.

## AN UNNAMED ROCK UNIT OF COMPARABLE AGE IN WRANGELL–ST. ELIAS NATIONAL PARK

Within the wide expanse of land known as Wrangell–St. Elias National Park in southeastern Alaska, there is an unnamed Late Cretaceous nonmarine sedimentary rock unit of comparable age to the Lower Cantwell Formation and Prince Creek Formation (map unit Ks, Richter, 1976; Kun, Richter et al., 2006). The exposures are very limited and are dominated by coarse conglomeratic units. Finer-grained sandstones and shales are minor components of the sequence. Small theropod and ornithopod dinosaur tracks were reported from these finer units (Fiorillo et al., 2012).

A small number of tree trunks preserved in upright growth position also occur (Figure 6.8). Only tracheids remain in the xylem of these fossil trees, indicating that the trees scattered across this part of the Cretaceous landscape were gymnosperms (Fiorillo et al., 2012). Other megafloral specimens include *Equisetites*, or

**FIGURE 6.7** Carbonized upright fossil tree trunk (shown by arrows) from the Chignik Formation of Aniakchak National Monument, Alaska. Hadrosaur footprints are associated with this specific stratigraphic horizon (not shown).

horsetails, and ferns (Fiorillo et al., 2012). Fossilized nitrogen-fixing structures were often found on the remains of the horsetails. The abundance of fossil ferns suggests that this landscape was a fern prairie or savanna (Fiorillo et al., 2012). Interestingly, rather than preserved as two-dimensional impressions, the fossil ferns within this rock unit are typically found preserved in three dimensions. The three-dimensional ferns and upright fossil tree trunks indicate rapid rates of sediment accumulation (Fiorillo et al., 2012) around the plants when they were living.

**FIGURE 6.8**  Upright fossil tree trunk from an unnamed Cretaceous rock unit in Wrangell–St. Elias National Park, Alaska.

## ANOTHER UNNAMED ROCK UNIT OF COMPARABLE AGE IN YUKON–CHARLEY RIVERS NATIONAL PRESERVE

Within a second and different unnamed fluvial sedimentary unit in the Yukon River drainage of east-central Alaska (Brabb and Churkin, 1969; Dover, 1994), there has also been a long history of the geologic story being told in part by plants. Spurr's (1898) original description of the sedimentary rocks around the Seventymile River, which drains into the Yukon River, is very brief. Without further details, he mentions that there is conglomerate associated with shale and sandstone containing abundant leaf impressions. Knowlton (Spurr, 1898) provides a list of fossil plants and attributes the age of this rock unit to the Tertiary, an age attribution that was carried through by subsequent workers until Hollick's work on the fossil plants of the region where Cretaceous floras were considered (Hollick, 1930). A more detailed discussion of the stratigraphy of this rock unit has been presented in Chapter 3.

A much more recent study of these rocks (Fiorillo et al., 2014) provided more detailed perspective on the paleoenvironment preserved within this unit. The pollen data indicate a landscape dominated by the gymnosperms *Metasequoia* and *Taxodiaceapollenites* (Fiorillo et al., 2014). All of the fossil wood in this region can be attributed to gymnosperms, while other megafloral remains include horsetails, *Metasequoia* leaves, and rare angiosperm leaves (Fiorillo et al., 2014). Included in the fossil tree remains is an exceptionally large gymnosperm trunk, 90 cm in diameter and preserved in three dimensions (Fiorillo et al., 2014). The floral composition in this part of Alaska during the Late Cretaceous suggests that the depositional environment was likely a *Taxodium*-dominated swamp or swampy margin of a low-gradient river system (Fiorillo et al., 2014).

## THE CIRCUM-ARCTIC VIEW

In addition to the abundant and growing dinosaur fossil record from the Cretaceous Arctic of Alaska, the extraordinary fossil plant record shows that the ancient Arctic was a highly productive circum-polar ecosystem surrounding a warm Arctic Ocean. The Late Cretaceous Arctic floras can be characterized as broad-leaved deciduous forests of varying diversity depending upon age and location, which gave way to conifer and fern-dominated communities in areas of high water table. The abundance and distribution of these fossil floras shows that this terrestrial ecosystem has no modern equivalent in the modern Arctic.

Relatively high rainfall, warm temperatures, and predominantly diffuse sunlight over much of a 24-hour day would have led to high-summer productivity (Herman et al., 2016). Cool, wet, and dark winters favored dieback, making these forests highly effective systems for sequestering carbon (Spicer et al., 1992). Supplying the moisture, the centrally positioned and warm Arctic Ocean may have been shrouded by a nearly permanent cloud cover (Spicer and Herman, 2010; Herman et al., 2016).

The fossil plant record of the Late Cretaceous Arctic is extraordinary, and while the discovery of fossil plants in these Cretaceous rocks occurred many decades before the discovery of dinosaurs, the details of the fossil plant record, when combined with the details of the fossil vertebrate record, illustrate beyond any doubt the remarkable complexity and heterogeneous nature of a truly unique ancient high-latitude terrestrial ecosystem in deep geologic time.

## REFERENCES

Beerling, D.J. and C.P. Osborne. 2002. Physiological ecology of Mesozoic polar forests in a high CO$_2$ environment. *Annals of Botany* 89:329–339.

Bell P.R. and E. Snively. 2008. Polar dinosaurs on parade: A review of dinosaur migration. *Alcheringa* 32:271–284.

Besse, J. and V. Courtillot. 1991. Revised and synthetic apparent polar wander paths of the African, Eurasian, North American and Indian plates, and true polar wander since 200 Ma. *Journal of Geophysical Research* 96:4029–4050.

Brabb, E.E. and M. Churkin, Jr. 1969. Geologic map of the Charley River Quadrangle, east-central Alaska. *United States Geological Survey Miscellaneous Investigations* Map 573, scale 1:250,000.

Brandlen, E. 2008. Paleoenvironmental reconstruction of the Late Cretaceous (Maastrichtian) Prince Creek Formation, near the Kikak–Tegoseak dinosaur quarry, North Slope, Alaska, *Unpublished MS thesis*, University of Alaska–Fairbanks, 225pp.

Brentnall, S.J., D.J. Beerling, C.P. Osborne, M. Harland, J.E. Francis, P.J. Valdes, and V.E. Wittig. 2005. Climatic and ecological determinants of leaf lifespan in polar forests of the high $CO_2$ Cretaceous 'greenhouse' world. *Global Change Biology* 11(12): 2177–2195.

Brongniart, A. 1825. Observations sur les végétaux fossiles renfermés dans les grès de Hoer en Scanie. *Annales des sciences naturelles* 4:200–224.

Cantrill, D.J. and I. Poole. 2012. *The Vegetation of Antarctica Through Geological Time*. Cambridge: Cambridge University Press.

Chaney, R.W. 1937. Age of the Cantwell Formation [abs.]. *Geological Society of America Proceedings* 1936:355–356.

Cockell, C., R. Corfield, N. Dise, N. Edwards, and N. Harris. 2008. *An Introduction to The Earth-Life System*. Cambridge, UK: Cambridge University Press.

Crane, P.R., P. Herendeen, and E.M. Friis. 2004. Fossils and plant phylogeny. *American Journal of Botany* 91:1683–1699.

Dall, W.H. 1896. Report on coal and lignite of Alaska. *United States Geological Survey Seventeenth Annual Report* 1:801–803.

Dover, J.H. 1994. Geology of part of east-central Alaska. In *The Geology of Alaska. The Geology of North America G-1*, eds. G. Plafker and H.C. Berg, 153–203. Boulder, CO: Geological Society of America.

Eldridge, G.H. 1900. A reconnaissance in the Sushitna basin and adjacent territory, Alaska, in 1898. *United States Geological Survey 20th Annual Report* 7:1–29.

Fiorillo, A.R. 2014. The dinosaurs of arctic Alaska. *Scientific American* 23:54–61.

Fiorillo, A.R., T.L. Adams, and Y. Kobayashi. 2012. New sedimentological, palaeobotanical, and dinosaur ichnological data on the palaeoecology of an unnamed Late Cretaceous rock unit in Wrangell–St. Elias National Park and Preserve, Alaska, USA. *Cretaceous Research* 37:291–299.

Fiorillo, A.R., F. Fanti, C. Hults, and S.T. Hasiotis. 2014. New ichnological, paleobotanical and detrital zircon data from an unnamed rock unit in Yukon–Charley Rivers National Preserve (Cretaceous: Alaska): Stratigraphic implications for the region. *Palaios* 29:16–26.

Fiorillo, A.R. and R.A. Gangloff. 2001. The caribou migration model for Arctic hadrosaurs (Ornithischia: Dinosauria): A reassessment. *Historical Biology* 15:323–334.

Fiorillo, A.R., P.J. McCarthy, P.P. Flaig, , E. Brandlen, D.W. Norton, P. Zippi, L. Jacobs, and R.A. Gangloff. 2010. Paleontology and paleoenvironmental interpretation of the Kikak-Tegoseak Quarry (Prince Creek Formation: Late Cretaceous), northern Alaska: A multi-disciplinary study of a high-latitude ceratopsian dinosaur bonebed. In *New Perspectives on Horned Dinosaurs*, eds. M.J. Ryan, B.J. Chinnery-Allgeier, and D.A. Eberth, 456–477. Bloomington: Indiana University Press.

Fiorillo, A.R. and J.T. Parrish. 2004. The first record of a Cretaceous dinosaur from western Alaska. *Cretaceous Research* 25:453–458.

Flaig, P.P., P.J. McCarthy, and A.R. Fiorillo. 2013. Anatomy, evolution and paleoenvironmental interpretation of an ancient Arctic coastal plain: Integrated paleopedology and palynology from the Upper Cretaceous (Maastrichtian) Prince Creek Formation,

North Slope, Alaska, USA. In *New Frontiers in Paleopedology and Terrestrial Paleoclimatology: Paleosols and Soil Surface Analogue Systems*, eds. S.G. Driese and L.C. Nordt, 179–230. North Slope, Alaska, USA: SEPM Special Publication 104.

Fredriksen, N.O. 1990. Pollen zonation and correlation of Maastrichtian marine beds and associated strata, Ocean Point Dinosaur Locality, North Slope, Alaska. *United States Geological Survey Bulletin* 1990-E, 1–24.

Greuter, W., J. McNeill, F.R. Barrie, H.M. Burdet, V. Demoulin, T.S. Filgueiras, D.H. Nicholson et al. 2000. *International Code of Botanical Nomenclature*. Konigstein: Koeltz Scientific Books.

Gryc, G., W.W. Patton, and T.G. Payne. 1951. Present stratigraphic nomenclature of northern Alaska. *Washington Academy of Sciences Journal* 41:159–167.

Harris, T.M. 1961. The fossil cycads. *Palaeontology* 4:313–323.

Herman, A.B. 2002. Late early–late Cretaceous floras of the North Pacific Region: Florogenesis and early angiosperm invasion. *Review of Palaeobotany and Palynology* 122:1–11.

Herman, A.B. 2004. The Cretaceous flora of the Kuk-Kaolak area, northern Alaska. *Stratigraphy and Geologic Correlation* 12:380–393.

Herman, A.B. 2007a. Comparative paleofloristics of the Albian-Early Paleocene in the Anadyr-Koryak and North Alaska Subregions, part 1: The Anadyr-Koryak Subregion. *Stratigraphy and Geological Correlation* 15:321–332.

Herman, A.B. 2007b. Comparative paleofloristics of the Albian-Early Paleocene in the Anadyr-Koryak and North Alaska Subregions, part 2: The North Alaska Subregion. *Stratigraphy and Geological Correlation* 15:373–384.

Herman, A.B. 2007c. Comparative paleofloristics of the Albian-Early Paleocene in the Anadyr-Koryak and North Alaska Subregions, Part 3: Comparison of floras and floristic changes across the Cretaceous–Tertiary boundary. *Stratigraphy and Geological Correlation* 15:516–524.

Herman, A.B. 2013. Albian-Paleocene flora of the North Pacific: Systematic composition, palaeofloristics and phyotstratigraphy. *Stratigraphy and Geological Correlation* 21:689–747.

Herman, A.B., R.A. Spicer, and T.E.V. Spicer. 2016. Environmental constraints on terrestrial vertebrate behavior and reproduction in the high Arctic of the Late Cretaceous. *Palaeogeography, Palaeoclimatology, Palaeoecology* 441:317–338.

Hisinger, W. 1841. *Lethaea suecia, supplem. secundi continuato*. Stockholm.

Hollick, A. 1930. The Upper Cretaceous floras of Alaska. *U.S. Geological Survey Professional Paper* 159:1–123.

Kimura, T. and S. Sekido. 1975. *Nilssoniocladus* n. gen. (Nilssoniaceae, n. fam.), newly found from the early Lower Cretaceous of Japan. *Palaeontographica Abteilung B* 153:111–118.

Krassilov, V.A. 1975. Climatic changes in Eastern Asia as indicated by fossil floras. II Late Cretaceous and Danian. *Palaeogeography, Palaeoclimatology, Palaeoecology* 17:157–172.

Langenheim, R.L., Jr., C.J. Smiley, and J. Gray. 1960. Cretaceous amber from the Arctic Coastal Plain of Alaska. *Bulletin of the Geological Society of America* 71:1345–1356.

Moffitt, F.H. and J.E. Pogue. 1915. The Broad Pass region, Alaska with sections on Quartenary deposits, igneous rocks, and glaciation. *United States Geological Survey Bulletin* 608:1–80.

Mull, C.G., D.W. Houseknecht, and K.J. Bird. 2003. Revised Cretaceous and Tertiary Stratigraphic Nomenclature in the Colville Basin, Northern Alaska. United States Geological Survey Professional Paper 1673. Version 1.0 http://pubs.usgs.gov/pp/p1673/index.html.

Osborne, C. P., D.L. Royer, and D.J. Beerling. 2004. Adaptive role of leaf habit in extinct polar forests. *International Forestry Review* 6:181–186.

Parrish, J.T. and R.A. Spicer. 1988. Late Cretaceous terrestrial vegetation: A near-polar temperature curve. *Geology* 16:22–25.

Peppe, D.J., J.M. Erickson, and L.J. Hickey. 2007. Fossil leaf species from the Fox Hills Formation (Upper Cretaceous: North Dakota, USA) and their paleogeographic significance. *Journal of Paleontology* 81:550–567.

Peralta-Medina, E. and H.J. Falcon-Lang. 2012. Cretaceous forest composition and productivity inferred from a global fossil wood database. *Geology* 40:219–222.

Pielou, E.C. 1988. *The World of Northern Evergreens*. Ithaca, NY: Comstock Publishing Associates.

Pott, C., S. McLoughlin, A. Lindström, S. Wu, and E.M. Friis. 2012. *Baikalophyllum lobatum* and *Rehezamites anisolobus*: Two seed plants with "Cycadophyte" foliage from the Early Cretaceous of eastern Asia. *International Journal of Plant Science* 173:192–208.

Read, J. and J. Francis. 1992. Responses of some Southern Hemisphere tree species to a prolonged dark period and their implications for high-latitude Cretaceous and Tertiary floras. *Palaeogeography, Palaeoclimatology, Palaeoecology* 99:271–290.

Retallack, G.J. and D.L. Dilcher. 1986. Cretaceous angiosperm invasion of North America. *Cretaceous Research* 7:227–252.

Richter, D.H. 1976. Geologic map of the Nabesna Quadrangle, Alaska. *United States Geological Survey Miscellaneous Investigations Series* Map I-932, 1:250,000.

Richter, D.H., C.C. Preller, K.A. Labay, and N.B. Shew. 2006. Geologic map of the Wrangell-Saint Elias National Park and Preserve, Alaska. *United States Geological Survey Scientific Investigations Series* Map SIM-2877, 1:350,000.

Royer, D.L., C.P. Osborne, and D.J. Beerling. 2003. Carbon loss by deciduous trees in a $CO_2$-rich ancient polar environment. *Nature* 424:60–62.

Savile, D.B.O. 1972. *Arctic adaptations in plants*. Canadian Department of Agriculture, Research Branch Monograph 6:1–81.

Schultz, J. 2005. *The Ecozones of The World*. 2nd edition. Berlin: Springer.

Shimer, G.T., J.A. Benowitz, P.W. Layer, P.J. McCarthy, C.L. Hanks, and M. Wartes. 2016. $^{40}Ar/^{39}Ar$ ages and geochemical characterization of Cretaceous bentonites in the Nanushuk, Seabee, Tuluvak, and Schrader Bluff formations, North Slope, Alaska. *Cretaceous Research* 57:325–341.

Smiley, C.J. 1966. Cretaceous floras of the Kuk River area, Alaska, stratigraphic and climatic interpretations. *Geological Society of America Bulletin* 77:1–14.

Smiley, C.J. 1969a. Floral zones and correlations of Cretaceous Kukpowruk and Corwin Formations, northwestern Alaska. *American Association of Petroleum Geologists Bulletin* 53:2079–2093.

Smiley, C.J. 1969b. Cretaceous Floras of the Chandler-Colville Region, Alaska, stratigraphy and preliminary floristics. *American Association of Petroleum Geologists Bulletin* 53:482–502.

Smiley, C.J. 1972a. Plant megafossil sequences, North Slope Cretaceous. *Geoscience and Man* 4:91–99.

Smiley, C.J. 1972b. Applicability of plant megafossil biostratigraphy to marine–nonmarine correlations: An example from the Cretaceous of Northern Alaska. *24th International Geoscience Congress* 7:413–421.

Spicer, R.A. 2003. Changing climate and biota. In *The Cretaceous World*, ed. P. Skelton, 85–162. Cambridge, UK: Cambridge University Press.

Spicer, R.A. and A.B. Herman. 1996. *Nilssoniocladus* in the Cretaceous Arctic: New species and biological insights. *Review of Palaeobotany and Palynology* 92:229–243.

Spicer, R.A. and A.B. Herman. 2001. The Albian-Cenomanian flora of the Kukpowruk River, western North Slope, Alaska: Stratigraphy and plant communities. *Cretaceous Research* 22:1–40.

Spicer, R.A. and A.B. Herman. 2010. The Late Cretaceous environment of the Arctic: A quantitative reassessment based on plant fossils. *Palaeogeography, Palaeoclimatology, Palaeoecology* 295:423–442.

Spicer, R.A., A.B. Herman, R. Amiot, and T.E.V. Spicer. 2016. Environmental adaptations and constraints on latest Cretaceous Arctic dinosaurs. *Global Geology* 19:241–254.

Spicer, R.A. and J.T. Parrish. 1990. Latest Cretaceous woods of the central North Slope, Alaska. *Palaeontology* 33:225–242.

Spicer, R.A., J.T. Parrish, and P.R. Grant. 1992. Evolution of vegetation and coal forming environments in the Late Cretaceous of North Slope Alaska: A model for polar coal deposition at times of global warmth. In *Controls on the Deposition of Cretaceous Coal*, eds. P.J. McCabe and J.T. Parrish, 177–192. Boulder, CO: Geological Society of America Special Publication 267.

Spicer, R.A., J.A. Wolfe, and D.J. Nichols. 1987. Alaska Cretaceous–Tertiary floras and Arctic Origins. *Palaeobiology* 13:73–83.

Spurr, J.E., 1898. Geology of the Yukon gold district, Alaska. *United States Geological Survey Eighteenth Annual Report* 3:87–392.

Stanton, T.V. and G.C. Martin. 1905. Mesozoic section on Cook Inlet, and Alaska Peninsula. *Geological Society of America Bulletin* 16:408–410.

Stewart, W.N. and G.W. Rothwell. 1993. *Paleobotany and The Evolution of Plants.* Cambridge: Cambridge University Press.

Tomsich, C.S., P.J. McCarthy, A.R. Fiorillo, D.B. Stone, J.A. Benowitz, and P.B. O'Sullivan. 2014. New zircon U-Pb ages for the lower Cantwell Formation: Implications for the Late Cretaceous paleoecology and paleoenvironment of the lower Cantwell Formation near Sable Mountain, Denali National Park and Preserve, central Alaska Range, USA. In *ICAM VI: Proceedings of the International Conference on Arctic Margins VI*, Fairbanks, Alaska, May 2011, eds. D.B. Stone, G.K. Grikurov, J.G. Clough, G.N. Oakey, and D.K. Thurston, 19–60. St. Petersburg: VSEGEI.

Tomsich, C.S., P.J. McCarthy, S.J. Fowell, and D. Sunderlin. 2010. Paleofloristic and paleoenvironmental information from a Late Cretaceous (Maastrichtian) flora of the Lower Cantwell Formation near Sable Mountain, Denali National Park, Alaska. *Palaeogeography, Palaeoclimatology, Palaeoecology* 295:389–408.

Witte, K.W., D.B. Stone, and C.G. Mull. 1987. Paleomagnetism, paleobotany, and paleogeography of the Cretaceous, North Slope, Alaska. In *Alaskan North Slope Geology*, eds. I.L. Tailleur and P. Weimer, 571–579. Sacramento, CA: Society of Economic Paleontologists and Mineralogists, Pacific Section.

Wolfe, J.A. and C. Wahrhaftig. 1970. The Cantwell formation of the central Alaska Range. *United States Geological Survey Bulletin* 1294-A:41–46.

# 7 Aspects of Paleobiology and Paleoecology

A common definition for the modern Arctic is that region north of the Arctic Circle (latitude 66° 33′ north) which is the approximate limit of the midnight sun (Figure 7.1). A second definition of the Arctic focuses on temperature, drawing the boundary as the isotherm following the average temperature in which the annual warmest month (July) is below 10°C (50°F). Other definitions place the Arctic boundary at the northernmost tree line, or the southernmost extent of sea ice, both of which mark abrupt biological and physical boundaries. Another criterion for defining the Arctic is the presence of continuous permafrost.

In applying the term Arctic to the Cretaceous, most of these definitions are unreasonable, but the limit of the midnight sun serves as an approximation of the Arctic even in the Cretaceous. Solar radiation and insolation broadly determine the distribution of climate and temperature. Therefore, the ecological distribution of animals, plants, and radiation varies as a function of latitude, so in effect all the criteria used to define the Arctic are correlated to some extent. Therefore the polar circles indeed demark unique areas of the globe that carry climatological and ecological significance even in deep time. Paleolatitudinal reconstructions are approximations rather than precise determinations of ancient geography. For convenience, I use the term "ancient Arctic" to define that region extending northward from the estimated Arctic Circle of the Cretaceous, and because this is an approximation, this region also includes latitudes within the modern sub-Arctic.

The rationale for the definition is rooted in an assumption based on the biotic responses to particular selection forces that are found within the modern Arctic ecosystem. The modern Arctic is an environment that selects for unique adaptations among its resident inhabitants, such as the decoupling of the usual diurnal cycle common in terrestrial vertebrates because of long periods of light and darkness (Blix, 2005). Cretaceous dinosaurs would have been subject to similar selection forces—an ancient Arctic environment.

Proverbial head scratching about how these animals survived in the ancient Arctic began quickly with the first discoveries of dinosaurs in Alaska. While his role in starting the great rush north for polar dinosaurs was previously overlooked by some, Charles Repenning wrote me in November, 2004 and generously provided me access to his early thoughts after he made those first determinations.

Reflecting back to when he opened the first box of dinosaur bones in November, 1983 (see Chapter 2), Repenning wrote:

> When I first opened the box (about 1′ × 1′ × 2′ in size, there were a lot of bones) of fossils that Shell sent me I was surprised. I had just published a report on Pliocene mammals from the Gubic Formation at nearby Ocean Point and expected more of the same.

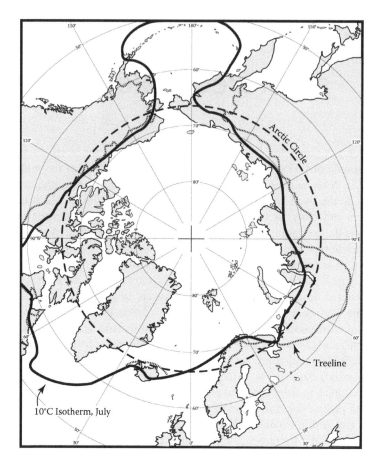

**FIGURE 7.1**   Polar projection comparing different demarcations for the modern Arctic—the Arctic Circle, the treeline, and the 10°C isotherm in July.

But they certainly weren't mammals and I quickly recognized the odd foot bones of a hadrosaur. The rest I didn't try to identify and assumed that they all belonged to the hadrosaur. I sent them all to Wann Langston because he had worked on hadrosaurs.

The modern climate up there is not compatible with reptiles of any sort and I worried a bit about the former climate.... I recognized that they had to be able to survive the long winter nights and paucity of vegetation during this time and I speculated that the hadrosaurs had fed on root mats of *Equisetum* in the shallow water of the Arctic Ocean and kept alive because the ocean was not frozen and would have been relatively warmer than the air. Plate tectonics indicates that the Arctic Ocean then was much smaller than now, and the locality was much closer to the modern North Pole then.

That satisfied me until Clemens (William) started digging up there and a whole dinosaur fauna began to emerge. Most of them were not adapted to eating *Equisetum* and my theory went to pot. (Repenning, 2004)

Pursuit of answers to the questions generated by the study of polar dinosaurs has involved many subsequent workers. Indeed, the specific issue of herbivory by hadrosaurs and ceratopsians near the shores of a shallow, warm Cretaceous Arctic Ocean has been considered in much more detail (Brinkman et al., 1998; Gangloff and Fiorillo, 2010; Gangloff, 2012). In the models proposed by these studies, herbivorous dinosaurs migrated seasonally from coastal areas to more inland areas to forage rather than migrate in a north–south manner. However, studies of stable oxygen isotopic signatures on the teeth of a variety of Alaskan dinosaurs show such speculation is not supported by geochemical data (Suarez et al., 2013). Stable isotopes have been recognized as valuable tools in reconstructing aspects of past climates. Specifically, isotopic signatures can be used to determine the isotopic composition of meteoric waters, that water derived from precipitation (Ludvigson et al., 1998; Ufnar et al., 2004; Suarez et al., 2011, 2013). As these waters are related to groundwater, the data gathered can be used in mass balance models to make inferences about the geochemistry of groundwater, the geochemistry of seawater, and aspects of temperature and humidity (Suarez et al., 2011) and when applied to fossil vertebrates, stable isotopes allow one to make inferences about diet (e.g., McKay, 2008; Fricke et al., 2009; Suarez et al., 2013).

In their study comparing the isotopic signature of Alaskan dinosaur teeth with those obtained from siderite (iron carbonate) nodules formed in the fossil soils of the Prince Creek Formation, Suarez and others (2013) reconstructed the paleohydrology of the Cretaceous Arctic coastal plain and made inferences about how dinosaurs consumed food resources. The teeth they examined are attributable to the predatory dinosaurs *Troodon*, *Dromaeosaurus*, a tyrannosaurid (= *Nanuqsaurus*, Chapter 4), and the herbivorous dinosaurs *Pachyrhinosaurus*, *Edmontosaurus*, and a hypsilophodontid (= thescelosaurid, Chapter 4), and the teeth came from three localities. Listing them in order from their paleogeographic position along a transect from more inland to more coastal, the three sites are Kikak-Tegoseak Quarry, Pediomys Point, and the Liscomb Bonebed.

This study (Suarez et al., 2013) determined that water ingested by most of the carnivorous dinosaurs tends to be more enriched in heavy oxygen ($^{18}O$) than for the herbivorous dinosaurs, which may reflect a difference in the proportion of drinking water to food water with high-protein content, a suggestion put forward by McKay (2008) to account for a similar isotopic pattern in a Pleistocene assemblage found in Natural Trap Cave, Wyoming. Suarez and others (2013) also determined that water ingested by *Pachyrhinosaurus* was more enriched in $^{18}O$ than for other herbivorous dinosaurs, and this pattern may be due to differences in drinking water source, consumed food water, or a combination of both.

With respect then to the idea that herbivorous Alaskan dinosaurs may have migrated to the coast during the Cretaceous Arctic winters to consume aquatic plants such as seagrass (Brinkman et al., 1998; Gangloff and Fiorillo, 2010; Gangloff, 2012), the isotopic data reported by Suarez and others (2013) do not support the model. Seawater is enriched with $^{18}O$, and the stable oxygen isotope values for seagrasses, because these plants live within the direct influence of seawater, should be similarly enriched in heavy oxygen. If animals are consuming seagrasses, the isotopic signature within their teeth should approach that signature reconstructed for seawater. The isotopic signals from *Edmontosaurus* and other non-*Pachyrhinosaurus* teeth do not show this convergence in values. Even with the highest $^{18}O$ values shown in this sample of teeth, if *Pachyrhinosaurus* consumed seagrass for a significant portion of the year, the $^{18}O$ values should be higher as well. Therefore, the geochemical signatures found in the teeth of these Arctic dinosaurs do not support a seagrass foraging model. Rather, *Pachyrhinosaurus* may have foraged for food out on the coastal plains away from riparian areas, while other herbivorous dinosaurs foraged for food in the riparian areas.

It is clear then from this example and those that follow that how these dinosaurs survived in the ancient Arctic and the dynamics between species and their ecosystem are driving issues in understanding the ancient polar world, and the only way to advance our understanding in a meaningful manner is to integrate data from across a variety of scientific disciplines.

## HAPPY HOMEBODIES OR WAYWARD WANDERERS?

The discovery of dinosaur remains at high latitudes proved problematic with respect to typical reptilian cold-blooded physiological models of a few decades ago. Hotton (1980), referring specifically to the dinosaurs found in the Yukon Territory of Canada (Rouse and Svrivastava, 1972), proposed a long-distance migration model whereby the animals migrated thousands of kilometers north and south annually on a journey tightly constrained by forage, temperature, and light conditions.

The abundant dinosaur remains found on the North Slope of Alaska in the 1980s reignited interest in the problem of high-latitude dinosaurs. Several additional workers (Brouwers et al., 1987; Parrish et al., 1987; Currie, 1989) embraced the hypothesis that these animals were long distance, seasonal migrants, with migratory ranges of several thousands of kilometers. These workers, using Caribou (*Rangifer tarandus*) as a modern analog, speculated that the dinosaurs migrated from northern Alaska to the more hospitable latitudes of southern Alberta during the winter months (Figure 7.2). In contrast, by considering duration of the freezing interval during the winter along with forage availability in the form of rhizomes and dried plant material, others have

**FIGURE 7.2** Caribou and Arctic hadrosaur models. Model (a) was proposed by early workers and suggests that caribou and their migration are an appropriate model for Arctic hadrosaur migration. Model (b) shows caribou migration behavior is not an appropriate model for Arctic hadrosaurs. Most modern workers accept model (b) (see text for discussion).

qualitatively argued that high-latitude dinosaurs were year-round residents (Brouwers et al., 1987; Parrish et al., 1987; Paul, 1988; Clemens and Nelms, 1993).

Patterns of migratory movements in vertebrates form a continuum of large-scale or small-scale movements. Small-scale movements lead from resource patch to resource patch related to exploiting a year-round food source, for example as is common among tropical birds (Rappole, 2013). Large-scale movements are recurrent seasonal and regular travels between the area where breeding takes place and an area of foraging where it does not. This pattern allows an organism to take advantage of favorable local conditions driven by strong seasonality in resource availability. It is this concept of large-scale movement that has been implicitly applied to the Alaskan dinosaurs (Parrish et al., 1987; Paul, 1988; Currie, 1989).

The use of caribou as a model for arctic dinosaurs is suspect for a couple of reasons. The first was the assumption that caribou moved in a linear north–south direction. The reality is that while some caribou herds do move in a general north–south manner, the patterns of caribou movement deviate, sometimes dramatically from north–south, as a function of local topography and resource availability (Figure 7.3). Moreover, caribou young must reach a certain size before they migrate. Because caribou are an important animal for rural communities throughout the Arctic, body size data are readily available, allowing age demographics to be determined. Examining data from three herds of caribou in Alaska (Teshekpuk, Western Arctic, and Northern Peninsula; Fiorillo and Gangloff, 2001) shows minimum individual size is roughly 60% adult mass and 85% adult length. The cell structure of the bones belonging to juvenile Alaskan hadrosaurs shows a pattern of textural shifts in rates of bone growth (Chapter 4) tied to seasonal changes that produce nutritional shortages (Chinsamy et al., 2012), indicating that the juveniles lived through several seasons. Thus, these hadrosaurs were more

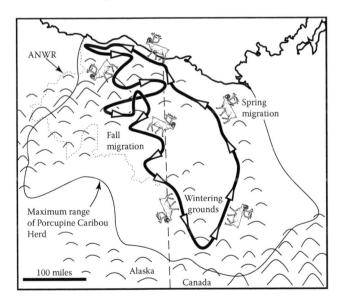

**FIGURE 7.3** Generalized map of the seasonal migration of the Porcupine Caribou Herd showing the non-linear pattern of seasonal movements.

than 1 year old at the time of death (Fiorillo and Gangloff, 2001; Gangloff and Fiorillo, 2010; Chinsamy et al., 2012), and in contrast to caribou, the relative size in these young hadrosaurs was only 11% adult mass and 23% adult length (Fiorillo and Gangloff, 2001). Biomechanically, Arctic hadrosaurs were probably too small to have migrated the thousands of miles required in the migratory model and instead they stayed in northern Alaska year round. In their study of the migration or overwintering question, Bell and Snively (2008) state that hadrosaurs and ceratopsians were likely not capable of the long-distance migrations invoked by Hotton (1980), even the adults.

Fricke and others (2009) measured the stable carbon and oxygen isotope ratios from the teeth of hadrosaurs recovered from correlative rock units in southern Alberta to New Mexico, assuming each sample of teeth represented a different population of hadrosaurs. The results showed that there was a large range in isotopic values due to differences in hadrosaur diet from place to place and in variations in the meteoric water they drank. Fricke and his colleagues concluded that these populations remained distinct and that hadrosaurs did not migrate. Though they did not sample Alaskan dinosaur teeth, their conclusions corroborate the idea that Alaskan dinosaurs did not migrate either.

## POPULATIONS

An important aspect of comparing the populations of modern Arctic animals to animal populations of the lower latitudes is that the northern populations of both herbivores and carnivores are prone to strong fluctuations (Figure 7.4). While the frequency of cycles can vary, the reason for the population cycles stems from the environmental extremes of the ecosystem. These environmental extremes produce strong stresses most clearly expressed in seasonal food resource availability. The seasonal aspect of these stresses makes them largely synchronized. Such stresses influence population dynamics, while in the lower latitudes those stresses are not necessarily synchronized and thus are not as dramatically reflected in population cyclicity.

The term "gregariousness" represents a spectrum of social interactions between various organisms, from complex social interactions derived from living and interacting together to simple interactions that result from coexistence due to environmental stress, such as drought-drawing species together at a water hole, or bounty, such as Grizzly bears (*Ursus arctos*) gathered during a salmon run or at a dead beached whale (Figure 7.5). One of the best sources of data for understanding populations and gregariousness in fossil vertebrates is from bonebeds. Bonebeds are units of strata that contain single large concentrations of fossilized bones and teeth. They are formed through mass death and very short-term depositional events. They reflect tightly constrained intervals of time and space and therefore provide snapshots of past ecosystems (Rogers et al., 2007).

Currently, no theropod bonebeds are known from the Alaskan Arctic and therefore discussion of gregariousness in Arctic theropods must be dubious. In southern Alberta, however, the tyrannosaur *Albertosaurus* may have been gregarious (Currie, 1998; Currie and Eberth, 2010). This notion stems from the occurrence of a unique bonebed in the Upper Cretaceous Horseshoe Canyon Formation. In this bonebed are the remains of at least 12 individuals of varying ontogenetic development *Albertosaurus* (Currie and Eberth, 2010; Eberth and Currie, 2010). While this bonebed reflects some aspect of gregarious behavior among these tyrannosaurs, it is

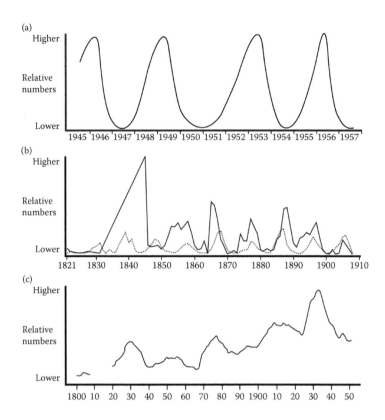

**FIGURE 7.4** Population cycles in Arctic animal populations showing patterns of significant population growth and decline. (a) Microtine rodents. (Redrawn from Pitelka, F.A. 1966. Some characteristics of microtine cycles in the Arctic. In *Arctic Biology*, ed. H.P. Hansen, 153–184. Corvallis, OR: Oregon State University Press, Corvallis.) (b) A comparison of Arctic hare (*Lepus arcticus*) populations (solid line) and Canada lynx (*Lynx canadensis*) populations (dotted line). (Redrawn from Seton, E.T. 1911. *The Arctic Prairies: A Canoe-journey of 2,000 Miles in Search of the Caribou; Being the Account of a Voyage to the Region North of Aylmer Lake*. Toronto: Briggs.) (c) Arctic Fox populations. (Redrawn from Vibe, C. 1967. *Meddelelser om Grønland* 170:1–227.)

unclear from the available data whether this site represents a group of cooperative hunting animals (Currie, 1998; Currie and Eberth, 2010) or a scenario such as a gathering of animals seeking high ground during a flood event (Eberth and Currie, 2010).

Although originally described from one locality, the Kikak-Tegoseak Quarry (see Chapter 4), *Nanuqsaurus*, the tyrannosaur of the Prince Creek Formation, is now known from at least three localities within the Late Cretaceous of northern Alaska (Fiorillo and Tykoski, 2014; Fiorillo et al., 2016). The limited material was recovered from a variety of subenvironments ranging from upper deltaic to lower coastal plain deposits, suggesting the species pursued its meals where it could find them and that its geographic range was not dictated by local environments. At this point, however, skeletal remains are scarce so there is no evidence to suggest that *Nanuqsaurus* was gregarious. Until a bonebed representing several individuals of this animal is discovered,

**FIGURE 7.5**    Grizzly bears fishing at Brooks Falls, Katmai National Park and Preserve, Alaska.

it is appropriate to consider that this Alaskan tyrannosaur was likely a solitary hunter and only social during breeding or unique feeding events, as commonly seen among modern solitary predatory animals such as Grizzly bears (*Ursus arctos*), which will gather to breed or to feed along stretches of river during salmon runs (Figure 7.5).

In contrast to the predators in the Alaskan Cretaceous ecosystem, there is ample evidence that herding occurred in the large herbivores *Pachyrhinosaurus* and *Edmontosaurus* (Fiorillo et al., 2010a,b; Gangloff and Fiorillo, 2010). The Kikak-Tegoseak Quarry (Chapter 4) is a bonebed that extends for at least 100 meters across an outcrop of the Prince Creek Formation (Fiorillo et al., 2010a,b). The most common bones from the site are those of *Pachyrhinosaurus perotorum*, an unusual horned dinosaur with a particularly distinguishing battering-ram-like face (Fiorillo and Tykoski, 2012). At least 11 individuals have been recognized from this one quarry thus far (Fiorillo and Tykoski, 2013). The number of individuals is based on one of the most common and easily identifiable elements found at the site, the occipital condyle, a round bony process found at the base of the skull where it articulates with the vertebral column (Figure 4.10). There is one occipital condyle per individual of *Pachyrhinosaurus perotorum*, and nine have been recovered. In addition, there are other individuals that can be added to the list because they are not of the appropriate size to match the size skull represented by these occipital condyles. For example, the rostral part of an immature *Pachyrhinosaurus perotorum* has also been recovered from this site (Fiorillo and Tykoski, 2013), and its reconstructed skull size is too small to be attributed to one of these condyles. The presence of so many individuals recovered from one quarry, with many more remaining in the ground, indicates that these animals lived in herds at least part of the year. Further, the range of the diameters of these condyles (Table 7.1) shows that this herd included younger and older generations (Fiorillo et al., 2010a, Fiorillo and Tykoski, 2013).

Two lines of evidence demonstrate hadrosaurs were gregarious in Alaska. The first is derived from bonebeds in the Prince Creek Formation that are dominated by the

**TABLE 7.1**

**Diameters of *Pachyrhinosaurus perotorum* Occipital Condyles Recovered and Prepared from the Kikak-Tegoseak Quarry**

| Specimen No. | Diameter (mm) |
| --- | --- |
| DMNH 22194 | 77.7 |
| DMNH 22195 | 80.8 |
| DMNH 22198 | 76.8 |
| DMNH 22257 | 79.5 |
| AK 539-V-01 | 65.1 |
| AK 539-V-12A | 74.7 |
| AK 539-V-28 | 73.5 |
| AK 539-V-29 | 76.7 |
| AK 539-V-11 | – |

*Source:* Data from Fiorillo, A.R. et al. 2010a. Paleontology and paleoenvironmental interpretation of the Kikak-Tegoseak Quarry (Prince Creek Formation: Late Cretaceous), northern Alaska: A multi-disciplinary study of a high-latitude ceratopsian dinosaur bonebed. In *New Perspectives on Horned Dinosaurs*, eds. M.J. Ryan, B.J. Chinnery-Allgeier, and D.A. Eberth, 456–477. Bloomington: Indiana University Press.

*Note:* A ninth condyle, though incomplete, is included here as a voucher for the existence of at least a ninth individual at this site. Additional individuals adding to the total of 11 are recognized from other bones and the clear size differences with individuals of these body sizes, such as the immature individual of *Pachyrhinosaurus perotorum* described by Fiorillo and Tykoski (2013). DMNH, Perot Museum of Nature and Science; AK, University of Alaska Museum.

remains of juvenile individuals of *Edmontosaurus* (Fiorillo et al., 2010b; Gangloff and Fiorillo, 2010). These bonebeds contain dozens of individuals (Fiorillo et al., 2010b) ranging in body size from 24% to 54% adult size (Gangloff and Fiorillo, 2010).

The second line of evidence comes from a remarkable tracksite in the Lower Cantwell Formation of Denali National Park. Here, on a single bedding plane, thousands of footprints are preserved, and the vast majority can be attributed to the ichnogenus *Hadrosauropodus*, tracks thought to have been left by hadrosaurs (Figure 7.6; Fiorillo et al., 2014a). Most hadrosaur tracks have skin impressions showing that the track surface is indeed the same one on which the dinosaurs walked (Fiorillo et al., 2014a).

Elephants are a commonly used modern analog for dinosaurs because they are the largest land animals of their day. Determining age profiles—the number of individuals in each age cohort—for elephant herds used to be challenging, for the obvious reason that the size of the animals prohibits safe close-up examination. However, a simple technique for estimating population age structure in elephant herds is to examine size differences in hind foot impressions left in the substrate (Western et al., 1983). Western and others (1983) showed a relationship between the size of tracks and the age of the elephants—simply that the largest tracks could be attributed to the most mature elephants, and progressively smaller tracks belonged to generations of

**FIGURE 7.6** Dinosaur tracksite in Denali National Park and Preserve, Alaska. This tracksite contains thousands of dinosaur footprints, the vast majority of which can be attributed to hadrosaurs, or duck-billed dinosaurs. The tracks range in size and include tracks attributable to very young hadrosaurs to fully grown adult hadrosaurs, indicating that high-latitude hadrosaurs lived in multigenerational herds. (Adapted from Fiorillo, A.R., S.T. Hasiotis, and Y. Kobayashi. 2014a. *Geology* 42:719–722.)

more immature individuals. Given this relationship in such a large modern animal, it is reasonable to apply the same technique to dinosaur track assemblages. Thus, the tracks at the large fossil tracksite in Denali National Park range in size from those of full-grown adult hadrosaurs to those of very small immature individuals (Fiorillo et al., 2014a; Figure 7.7). The most common animals within the herd were the adults, represented by 84% of the tracks found, followed in frequency by the very small individuals, which are represented by approximately 13% of the tracks found at the site. A third set, approximately 3% of the tracks at this site, represents animals larger than the youngest animals but decidedly smaller than the adults (Fiorillo et al., 2014a). The frequency of this mid-range group is very small, suggesting that hadrosaurs grew quickly through this size range before achieving adulthood.

## NICHE PARTITIONING

Like many who grew up under the influence of Theodore Seuss Geisel—known more commonly as Dr. Seuss—my first introduction to the concept of niche partitioning in ecosystems came from *On Beyond Zebra!* and the problems of the Nutches (Geisel, 1955). Ecological niche theory according to Dr. Seuss went like this:

> And NUH is the letter I use to spell Nutches,
> Who live in small caves, known as Niches, for hutches.
> These Nutches have troubles, the biggest of which is

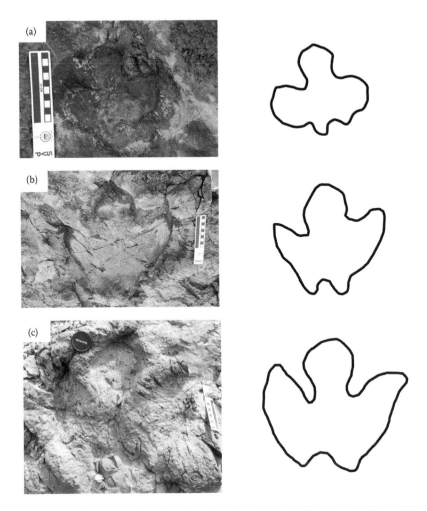

**FIGURE 7.7** Hadrosaur tracks of various sizes from the tracksite shown in Figure 7.6: (a) is one of the smallest tracks found and corresponds to a very young individual (Stage 1 of Fiorillo et al., 2014a); (b) is a track made by an immature hadrosaur and can be attitributed to Stage 3 (Fiorillo et al., 2014a); and (c) is a very large hadrosaur track attributable to an adult individual (Stage 4 of Fiorillo et al., 2014a).

The fact there are many more Nutches than Niches.
Each Nutch in a Nich knows that some other Nutch
Would like to move into his Nich very much
So each Nutch in a Nich has to watch that small Nich
Or Nutches who haven't got Niches will snitch.

Why organisms do what they do, how they live, what they eat, how they interact and coexist, and understanding the determinants of an organism's abundance and distribution are central issues in understanding an organism's ecological niche. Nearly a century ago it was understood that two species coexist by differing in their

lifestyles and traits (Grinnell, 1917). Patterns of food acquisition are basic determinants in the behavior of organisms now, and by extension, in the past.

What, then, were the patterns of food acquisition in Alaskan dinosaurs? More specifically, is there evidence of food resource partitioning among these animals, particularly among the large herbivores? How do dietary patterns in Cretaceous Alaska
compare to elsewhere in ancient North America? Were the dietary preferences of
high-latitude-dwelling hadrosaurs different from those found in lower latitudes?

Evidence for niche partitioning is found in the rich dinosaur track record of the
Lower Cantwell Formation of Denali National Park. Three very different types of
theropod tracks are found within this rock unit: a four-toed variety, a smallish three-
toed type, and a two-toed type (Figure 7.8). These three shapes of tracks are clues to
the identities of the trackmakers. The four-toed tracks are attributed to therizinosaurs

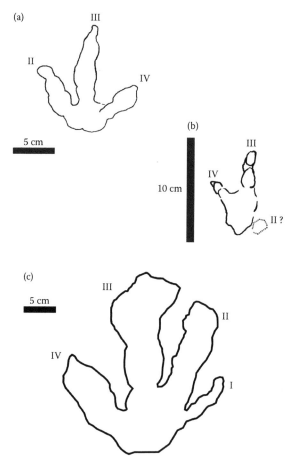

**FIGURE 7.8** Illustration of theropod track diversity now recorded from Denali National
Park and Preserve, Alaska. (a) A primitive three-toed theropod track. (b) A derived two-toed
theropod track. (Adapted from Fiorillo, A.R. et al. 2014b. *New Mexico Museum of Natural
History and Science Bulletin* 62:429–439.) (c) A four-toed therizinosaur track. (Adapted from
Fiorillo, A.R. and T.L. Adams. 2012. *Palaios* 27:395–400.)

(Fiorillo and Adams, 2012), unusual theropods because they are considered herbivorous, or at most omnivorous (Clark et al., 2004). Thus, that trackmaker would not have been in competition for food resources with carnivorous theropods from Alaska.

The didactyl (two-toed) and tridactyl (three-toed) tracks (Fiorillo et al., 2014b) indicate clear morphological distinction of the feet, reflecting locomotor and therefore behavioral differences between the trackmakers, implying additional ecological separation among theropod groups. Size, speed, agility, and weaponry of the carnivore, all of which can be inferred from tracks, help define the features of its niche.

Tridactyl theropod tracks from the Lower Cantwell Formation represent the plesiomorphic, or basal and less-specialized foot condition of Neotheropoda (coelophysoids, a group of slender carnivorous dinosaurs defined in part by characters within the front of the skull, plus birds; sensu Sereno, 1998). In contrast, the similarly sized didactyl theropod tracks belong to the ichnofamily Dromaeopodidae, with trackmakers being deinonychosaurs, dinosaurs famous for having a large sickle-like claw on pedal digit II. This large claw is evidence these animals were active predators rather than scavengers and that they were cooperative hunters (Ostrom, 1969), an interpretation not universally accepted (Roach and Brinkman, 2007). In addition, arguments have been made from both skeletal remains (Maxwell and Ostrom, 1995) and footprints (Li et al., 2008) that deinonychosaurs probably moved in groups. Further, it is reasonable to expect they hunted cooperatively at least part of the time. The energetic tradeoff between cooperative hunting and solitary hunting is simply whether the costs of cooperative hunting are less than the costs of sharing the kill. If the costs of the cooperative hunt outweigh the costs of sharing the meat, then solitary hunting strategies are selected for (Ricklefs and Miller, 2000). The benefit of cooperative hunting is that pack hunters can bring down larger prey items than solitary hunters. The derived claw on digit II of deinonychosaurs seems likely to reflect a grappler or slasher and scansorial hunting style (sensu Van Valkenburgh and Molnar, 2002; Manning et al., 2006, 2009; Fowler et al., 2011). Typical theropod foot impressions were left by animals that hunted much smaller prey than deinonychosaurs of comparable body size (Van Valkenburgh and Molnar, 2002). The large, slashing toe and possible cooperative hunting style of deinonychosaurs allowed them to target larger prey than their less-specialized counterparts. Thus, tracks provide evidence of complexity within the theropod guild of this ancient high-latitude terrestrial ecosystem.

Among herbivores, pits and scratches viewed with traditional light and scanning electron microscopes are features of microwear found on enamel, the former feature being associated with processing harder food items such as tough, woody, or fibrous vegetation, and the latter feature associated with softer foods such as leaves. While many workers define "pits" as features having length to width ratios ranging from less than 2:1, some authors have extended the ratio to 4:1. "Scratches" are certainly features with ratios exceeding 4:1. As shown by modern animals, the numbers of pits and scratches can change with variation in diet, for instance with seasonal change in available plant food.

In examining microwear in sympatric *Diplodocus* and *Camarasaurus*, Jurassic sauropods from western North America, niche partitioning was demonstrated because the teeth of these two dinosaurs have distinctively different wear patterns, indicating differences in diet. *Diplodocus* had a relatively softer diet, such as ginkgo leaves, and *Camarasaurus* a coarser diet, such as cycad foliage (Fiorillo, 1998). Similarly, Goswami and others (2005) used microwear to infer diet in two species of cynodonts (an early evolutionary branch that includes mammals) and a prosauropod dinosaur from the Triassic of Madagascar. Their study also showed aspects of difference in food use patterns in these animals, specifically that prosauropods likely ate softer foods than did the cynodonts.

The two most common very large herbivores in the Cretaceous of Alaska that are known from fossil bones are the hadrosaur *Edmontosaurus* sp. (Gangloff and Fiorillo, 2010; see Chapter 4 for discussion of this dinosaur) and the ceratopsian *Pachyrhinosaurus perotorum* (Fiorillo and Tykoski, 2012). Adult hadrosaurs weighed up to several tons, while adult ceratopsians weighed perhaps a couple of tons.

In addition to herbivorous diet and large body sizes, these animals had teeth that loosely resembled each other and formed tooth batteries that created large grinding and shearing surfaces for processing food. The chewing process in hadrosaurs and ceratopsians was very different, however. Ceratopsian jaws are more robust than hadrosaur jaws, which means that the former used powerful jaw motion to shear tough vegetation. In contrast, hadrosaurs had a weaker bite force and some degree of grinding between upper and lower teeth (Cuthbertson et al., 2012). Because of this difference in jaw mechanics, one can predict that microwear on the teeth of Alaskan *Edmontosaurus* and *Pachyrhinosaurus* would show different patterns of wear.

This is exactly the case. *Pachyrhinosaurus* enamel shows a pattern of heavy wear that includes abundant scratches of various thicknesses and some pitting indicative of hard food items (Figure 7.9). In contrast, in Alaskan hadrosaur teeth, the lower density of scratches as well as low frequency of pitting suggests this animal consumed softer food items than those consumed by *Pachyrhinosaurus* (Figure 7.10). These two large herbivores separated their niches by eating different plants.

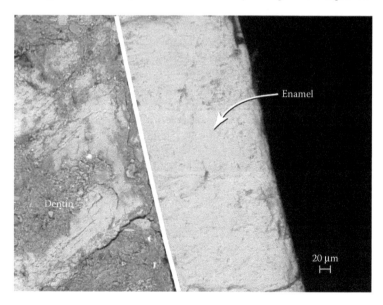

**FIGURE 7.9** Photograph of microwear pattern on the tooth of the Arctic ceratopsian dinosaur *Pachyrhinosaurus perotorum* (DMNH 22156). Notice the heavy scratching, and particularly the pitting, on the enamel.

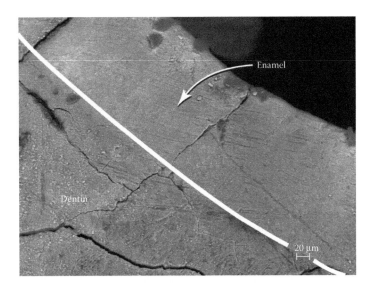

**FIGURE 7.10**    Photograph of microwear pattern on an Arctic *Edmontosaurus* tooth (DMNH 22143). Compare this tooth to the ceratopsian tooth in Figure 7.9 and notice the absence of pitting in this hadrosaur tooth, which indicates a diet of softer vegetation. The white line is drawn to show the boundary between the enamel and dentin tissues.

Of the five bonebeds known the Prince Creek Formation, only one is dominated by the remains of *Pachyrhinosaurus*, the others by hadrosaurs. Four of the five are from the more distal areas of the depositional basin with delta plain facies, while the fifth bonebed is from the more proximal part of the basin and represents a better-drained coastal plain.

The bonebeds in the distal areas are dominated by bones of *Edmontosaurus* sp. (see Chapter 4 for taxonomic discussion), while the more proximal bonebed is dominated by the remains of *Pachyrhinosaurus perotorum*. The distribution of these bonebeds and sedimentological facies suggest that *Pachyrhinosaurus* preferred more upland environments, which were populated by conifers (Flaig et al., 2013), while *Edmontosaurus* preferred more lowland, deltaic environments, which were populated by lowland trees, shrubs, and herbs (Flaig et al., 2013). Assuming, as is likely, that different kinds of plants inhabited these two environments, the differences in wear patterns on the teeth would reflect differences in the plant communities in which they foraged.

The presence of *Pachyrhinosaurus* in the better-drained upland areas contrasts with reports of large ceratopsian remains found in the lower latitudes. Similarly, hadrosaur remains have generally been ubiquitous in the lower latitudes but seem to be preferentially preserved in the more poorly drained low areas. These contrasting distributions were likely driven by changes in the distribution of optimal forage resulting from the pronounced seasonality of the polar terrestrial ecosystem.

During the Late Cretaceous, hadrosaurs had an enormous geographic range extending from northern Alaska (Gangloff and Fiorillo, 2010) to Baja, Mexico (Weishampel et al., 2004). Though a seemingly large geographic distribution, it does have modern analogs. The modern mountain sheep of North America have a comparable latitudinal range. These sheep range from the Brooks Range in Alaska, extending south through western Canada and the United States, to northwestern Mexico (Valdez and Krausman, 1999).

In that geographic range, a great variety of plant species is available for food. While sheep eat a broad spectrum of plants across that range, the northern Dall and Stone's Sheep consume more lichens and mosses than other North American sheep species (Nichols and Bunnell, 1999). Although lichens have a high digestibility, they have low-nutritional value. However, because of the short growing season in the higher latitudes, sheep in these northern environments are forced to consume food items that are less than optimal for their needs.

From observations of diet in North American mountain sheep across their modern range, a wide range in microwear patterns on the teeth of hadrosaurs would be expected across their Cretaceous range. In fact, in a preliminary study such was not the case. Instead, the wear patterns were remarkably similar (Fiorillo, 2011). While it may be that these northern hadrosaurs were indeed eating a wider array of food items than their more southerly counterparts, the textures of those foods across latitudes were similar, and the similar textures masked dietary differences recorded in microwear.

## COPROLITES

The naturalist Martin Lister (1678) has typically been credited with the first descriptions of fossil fecal material, making observations on molluscan fecal pellets (El-Baz, 1968; Thulborn, 1991; Northwood, 2005), though more recently it has been argued that the first published proper scientific account of fossil fecal material with a figure was in 1699 by Edward Lhwyd (Duffin, 2012). It wasn't until the first half of the 19th century that reports of coprolites began to populate the literature (e.g., Mantell, 1822; Buckland, 1829; Dekay, 1830), and the now familiar term "coprolite" was coined by Buckland (1829). These studies as well as the subsequent work over the next 100+ years provided only somewhat cursory examination of the paleobiological significance of coprolites. It has only been over the last few decades that a deeper appreciation of the potential for paleobiological information from coprolites has arisen, particularly with respect to fossil terrestrial reptiles such as dinosaurs, where meaningful insights are now gained about the nature of food webs (e.g., Thulborn, 1991; Hunt, 1992; Hunt et al., 1994; Chin and Gill, 1996; Chin et al., 1998; Chin, 2002, 2007; Hollocher et al., 2005; Northwood, 2005).

In the Lower Cantwell Formation of Denali National Park, in addition to the fossil fish traces and numerous footprints (Chapter 5), there are now several occurrences of coprolites (Figure 7.11). These specimens are considered as such because the gross morphology, the macroscopic surface features (pustulate surfaces), and the lack of clastic grains contained within are characteristics seen in examples of modern feces. Carbonized plant material can be seen on the surfaces of some of these specimens. Geochemical analysis of specimens has also shown elements such as carbon, calcium, manganese, phosphorus, potassium, and sulfur (Figure 7.12). These elements are commonly found in the manure of livestock (Pettygrove et al., 2010; Manitoba Agriculture, Food and Rural Development, 2015).

But what animal produced the coprolites? In their extensive reviews of spiral coprolites, Williams (1972) and McAllister (1985) pointed out the range of aquatic

**FIGURE 7.11**    Coprolite, likely from a hadrosaur, from Denali National Park and Preserve, Alaska.

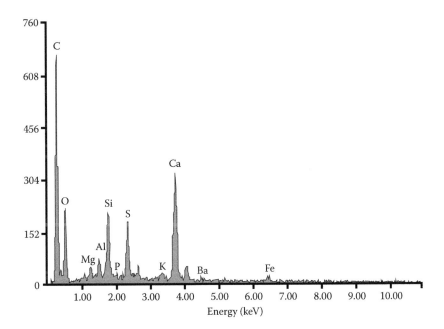

**FIGURE 7.12**    X-ray fluorescence (XRF) elemental analysis of a coprolite from the Lower Cantwell Formation of Denali National Park, Alaska. Note the presence of carbon, calcium, manganese, phosphorus, potassium, and sulfur—all elements found in the manure of livestock.

chondricthyans and osteichthyans that produce spiral fecal material. In most examples, the spiral structure observed as an external feature is also reflected as part of an internal structure. However, no such internal structure was observed in these coprolites from the Lower Cantwell Formation. Some teleosts have internally structureless fecal material (Williams, 1972); there are no records of teleosts of coprolites of this size, likely ruling out attribution of these coprolites to teleosts.

Jain (1983) described a variety of Triassic spiral coprolites from the Maleri Formation of India. As a review within that study, it was pointed out that though there are rare reports of spiral coprolites that have been attributed to some fossil tetrapods, no modern tetrapod makes such fecal material. Because these coprolites are clearly not attributable to fishes, and given the abundance of dinosaur footprints from this rock unit (Chapter 5), it is most likely that these coprolites were produced by dinosaurs. Given the occurrence of carbonized plant material in some specimens, as well as the shape (Figure 7.12), these coprolites are most likely from herbivorous dinosaurs. Further, given the abundance of *Hadrosauropodus* tracks in this rock unit, most of these coprolites were probably produced by hadrosaurs.

## ADAPTATION

Biological adaptation can be defined as the state of evolutionary fitness an organism has achieved, allowing it to successfully compete with other species and to tolerate the physical environment. Adaptation has been invoked for many taxa across a variety of environmental settings and throughout time and space (Novacek, 1996). Thus, rather than address adaption as a function of polar faunal provinciality, this discussion remains focused on adaptation.

There are a plethora of adaptative responses by organisms to the seasonal polar environment with its marked non-growing season as the result of a profound shift in light regime. Modern vertebrates deal with these conditions through behavioral, physiological, or phenotypical changes in contrast to related forms in less seasonal regimes. Evidence shows that dinosaurian taxa may have responded similarly to these conditions.

Blix (2005) argued that one adaptation to the Arctic is unique: the ability of organisms to decouple diurnal responses to environmental stimuli. Most tetrapods, for example, adjust their metabolic rate to the ambient temperature. When temperatures are warmer for a given animal, bone growth rates increase, and when ambient temperatures cool, bone growth rates correspondingly decrease. In this case, metabolic responses to the physical environment are expressed in the histologic structure of bone, which is observable in fossils.

Histological sections were made from limb bones of Alaskan hadrosaurs and compared to similar histological sections made from individuals of *Edmontosaurus* from the Upper Cretaceous Horseshoe Canyon Formation of southern Alberta, a rock unit that represents a more temperate paleoclimate. Examination of the polar forms of this dinosaur showed periodic textural shifts in all bones (Figure 7.13). These shifts were in the form of alternating cycles in the deposition of faster growing, more energetically demanding reticular fibrolamellar bone versus slower-formed circumferential fibrolamellar bone. In contrast, textural switches were not consistently observed in the bones of temperate *Edmontosaurus*. These alternating

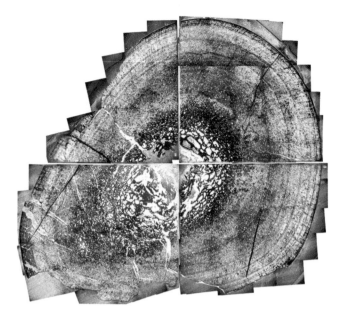

**FIGURE 7.13** Histological section made from Arctic hadrosaur limb bone from the Prince Creek Formation (DMNH 22552). This bone shows periodic textural shifts in bone growth.

deposits of bone in the polar dinosaurs are consistent with seasonally increased energy demands, or more specifically, the onset of the arctic winter with its subsequent loss of foliage caused the shift in bone deposition from reticular to circumferential fibrolamellar bone.

Similarly, Erickson and Druckenmiller (2011) examined limb bones of the ceratopsian *Pachyrhinosaurus perotorum*. In their study, they also found a pattern of conspicuous banding in the deposition of bone tissues, a feature not found in southern ceratopsids. Both of these studies show the largest of the herbivores found in this ecosystem were well adapted for life in a highly seasonal environment.

The microwear patterns on the teeth of similar sympatric theropods show identical wear patterns, and thus no apparent differentation of diet to reduce competition for food resources (Fiorillo, 2008). *Troodon*, a smallish theropod found throughout western North America, is normally considered an uncommon or rare faunal component in southern Canada and the western United States, but *Troodon* is the most common theropod known from the Prince Creek Formation (Fiorillo and Gangloff, 2000; Fiorillo, 2008). Fiorillo and Gangloff (2000) suggested that the abundance of *Troodon* in northern Alaska was because *Troodon* had a preadaptive character— large orbits (and by inference, large eyes), which gave the animals a competitive advantage in the domain of low-angle light. This advantage seems to have translated to an increased supply of food for *Troodon* individuals, resulting in an increase in average body size (Fiorillo, 2008; Figure 7.14).

Rather than increased body size through an adaptive advantage, the estimated body size of the apex predator, the tyrannosaurid *Nanuqsaurus*, shows this animal was of smaller size than coeval and related tyrannosaurids from the lower latitudes

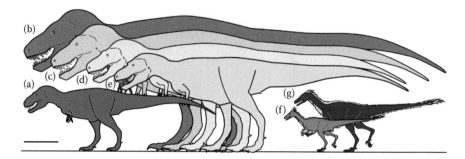

**FIGURE 7.14**  Silhouettes comparing representative theropods. (a) *Nanuqsaurus hoglundi*, based on holotype, DMNH 21461. (b) *Tyrannosaurus rex*, based on FMNH PR2081. (c) *Tyrannosaurus rex*, based on AMNH 5027. (d) *Daspletosaurus torosus*, based on FMNH PR308. (e) *Albertosaurus sarcophagus*, based on TMP 81.10.1. (f) *Troodon formosus*, lower latitude individual based on multiple sources and size estimates. (g) *Troodon* sp., North Slope individual based on extrapolation from measurements of multiple dental specimens. Scale bar equals 1 m. Notice the body size increase in northern forms of *Troodon* and the body size decrease in northern forms of tyrannosaurs, suggesting that polar ecosystems selected for an optimal body size within theropods. (Figure from Fiorillo, A.R. and R.S. Tykoski. 2014. *PLoS ONE* 9(3):e91287. doi:10.1371/journal.pone.0091287.)

(Fiorillo and Tykoski, 2014; Figure 7.14). The smaller body size of *Nanuqsaurus* may reflect lower seasonal carrying capacity due to the widely varying light regime affecting biological productivity. In that environment, the phenotypic response by the apex predator was to reduce body mass (Fiorillo and Tykoski, 2014).

Based on the dental data, this Alaskan *Troodon* seems to have reached almost 4 meters in length, compared to the southern forms which reached lengths of approximately 2.5 meters. In contrast, *Nanuqsaurus* reached a length of approximately 6 meters, about half the body length of a full-sized *Tyrannosaurus rex*. The increase in body size by one theropod taxon (*Troodon*) and the reduction in body size by a second theropod taxon (*Nanuqsaurus*) indicates an external ecological pressure for optimal body size. That pressure was likely dictated by the effective net biological productivity during the growing season.

Though these theropods converged in body size, there is evidence that their niches remained separate. Derived tyrannosaurs in general are recognized as having enhanced olfaction (Zelenitsky et al., 2009) and as a derived tyrannosaur, *Nanuqsaurus* likely had a similar sensory advantage. In contrast, the sympatric *Troodon* had the adaptative advantage of larger eyes in the highly seasonal physical environment, a feature that selected for larger body size. Therefore, niche separation within the predatory dinosaurs of the ancient polar ecosystem, it seems, was maintained by reliance on different senses to acquire food items.

## WHAT COLOR WERE ARCTIC DINOSAURS?

A quick perusal of dinosaur artwork shows, among other things, a decided shift in the coloration of dinosaurs by artists through time. Older works by artists such as Charles R. Knight and Rudolph F. Zallinger tended to portray dinosaurs in drab tones, while more contemporary artists have tended to use bold colors. The use of drab tones was driven by comparisons with large modern animals such as elephants, but the broad acceptance of the intimate relationship of birds to dinosaurs has changed the discussion.

The use of mammals as a model for the coloration of dinosaurs was flawed to begin with. Consider, for example, why a human hunter can wear blaze orange clothing while deer hunting. It is because most mammals are color blind, hence the use of color is not a tool for communication within many mammal species. In contrast, birds see in color, as can many non-avian reptiles as well, suggesting that color vision is the primitive condition within this clade. Color-seeing species use color to communicate in mating and marking territory, so a better understanding of dinosaurian and avian phylogeny has provided a new perspective on what the coloration patterns of fossil birds and non-avian dinosaurs may have been.

Remarkable preservation of fossil feathers at the famous localities of the Middle Eocene Messel Oil Shale in Germany and the Jurassic and Cretaceous deposits of Liaoning Province, China, have provided us direct fossil evidence for color patterning in ancient birds and dinosaurs (Li et al., 2010; Vinther et al., 2010; Zhang et al., 2010). These exquisite new discoveries show that these animals had plumage patterns of light and dark, and there is even evidence of iridescent blue, black, green, and coppery sheens in some plumages (Li et al., 2010; Vinther et al., 2010; Zhang et al., 2010). It now appears that much like modern birds, coloration patterns of dinosaurs and ancient birds probably contributed to camouflage, thermoregulation, and behavior patterns.

Thanks to these new discoveries it is entirely pertinent, then, to consider what color arctic dinosaurs might have been. Coloration of organisms can be vital across a variety of biota. With modern Arctic plants, for example, a trend toward darker colors helps warm the plants even when plants are under thin snow cover (Savile, 1972). An animal's coloration is often an important basis of selection relative to predation, communication, and metabolism (Hamilton, 1973). Are there generalizations about coloration in modern animals that can be used to speculate about coloration in Arctic dinosaurs?

In the modern world, Gloger's Rule, named for the zoologist Constantin Wilhelm Lambert Gloger, states that vertebrates found near the equator tend to be more darkly colored than those found in higher latitudes. Gloger first remarked upon this phenomenon in 1833, specifically related to climate and color in avian plumage. The rule has been extended more broadly to pertain to endotherms in general. The pattern seems to be reversed in some ecotherms, however. The Wood frog (*Rana sylvatica*), for example, has a geographic range from North Carolina to interior Alaska. Northern forms of this frog are more vividly marked with dark coloration compared to their lower latitude contemporaries (Hodge, 1976). Is there an adaptative advantage of a lighter color versus a darker color in higher latitude animals?

The classic interpretation is that white coloration is to reflect sunlight, while black coloration is beneficial in absorbing it (Figure 7.15). In other words, white is a protection against excessive heat buildup. But if that were true, Hamilton (1973) pointed out that there should be white animals in the deserts instead of the dark-colored animals that are found there. Basic thermoregulatory adaptation of birds and mammals to the Arctic environment is to increase insulation thickness because the insulation protects the animal from excessive cold while at rest (Scholander et al., 1950).

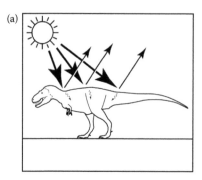 

**FIGURE 7.15** Illustration of the relationship between color and energy absorption in dinosaurs. A light-colored animal reflects a significant amount of energy compared to the energy absorbed (a), while with a dark-colored animal more energy tends to be absorbed (b).

Animals such as Arctic lemmings (*Dicrostonyx torquatus*), Snowshoe hares (*Lepus americanus*), Short-tailed weasels or ermine (*Mustela erminea*), and Willow ptarmigan (*Lagopus lagopus*), which are white in the winter, change to brown during summer, when closely related animals in other areas do not change color. Snowshoe hares tend to remain still when sensing danger, and ptarmigan stick to the background that matches their phase of molt (Hamilton, 1973). These behaviors suggest that camouflage plays a role in coloration.

However, as Marchand (1996) points out, given that Short-tailed and Least weasels spend their winter time hunting under the surface of the snow while in their white phase, why is white important? Further, if white was important as a cryptic coloration phenomenon, then the red fox and the fisher should turn white as they hunt above the snow (Marchand, 1996). In the Pribilof Islands, as well as many coastal areas of Alaska and Canada, the blue phase of the Arctic fox is most common, rather than the white. If white is advantageous for snowy, winter camouflage, then why do these animals remain slate blue (Marchand, 1996)?

In a classic study, Hamilton and Heppner (1967) dyed white zebra finches black and placed the study birds in a chamber where a light could serve as a sun and turn on and off. With the light on, the dyed birds metabolized at a considerably lower rate than the white birds. It seems then that animal coloration plays little to no role in the loss of heat; rather, coloration only contributes to the relative rate of heat gain (Hamilton, 1973). However, under field conditions that relationship is a bit more complicated. In a study of two subspecies of *Lagopus lagopus*, the European red grouse (*L. l. scoticus*) and the Scandinavian willow grouse (*L. l. lagopus*), Ward and others (2007) measured the thermal consequences between these two birds. The Red grouse did not change color, while the Willow grouse did change from a dark plumage to a white one. In still air, heat gained was greater in birds with a dark plumage, but in windy conditions there was no measurable difference in heat gain (Ward et al., 2007). Their study implies that there is a tradeoff between camouflage and thermal benefits to animals based on white coloration.

So why aren't there more black or dark-colored animals in the north? Hamilton (1973) studied 27 species of all-white and 39 species of all-black birds in western North America and showed that the insulation of all-white birds was superior to all-black birds. This was expressed in the patterns of roosting behavior; all-white birds roosted in open areas, all-black birds roosted under cover. Hamilton (1973) suggested that white coloration minimizes heat exchange between an organism and environment. The cause of whiteness is the absence of melanin (Marchand, 1996). Melanocytes produce melanin, and in their study of *Mustela erminea* from New Zealand, King and Moody (1982) showed that cold climates inhibit the synthesis of melanocyte-stimulating hormones that create brown hair in these animals. Because the white hair of mammals and the white feathers of birds are hollow, incoming light is scattered and reflected rather than absorbed (Marchand, 1996), providing the physical reason for white being a better insulator in Arctic animals.

Thus, there are conflicting data for resolving why animals turn white in the Polar Regions, and no consensus as to why modern animals seasonally turn white. However, regardless of whether animals change color for camouflage or for thermal regulation, ptarmigan change color. Their feathers are good insulation and they are replaced throughout the year. It is also clear that birds are dinosaurs (Chapter 4, Figure 4.1). The southern range for ptarmigan in North America is approximately the border of Canada and the continental United States, with outlier populations occurring even further south in the rocky tundra of the states of Colorado and Wyoming (Sibley, 2000). Because the temperature reconstruction for the ancient Arctic (Chapter 8) is approximately the same as the southern extent of ptarmigan in North America, and temperature—specifically low temperature—is the inhibitor for the production of melanin, it is intriguing to think that dinosaurs such as *Troodon* and *Nanuqsaurus*, predatory dinosaurs that hunted above ground throughout the year, were indeed white at times during the ancient Arctic winter.

## SPECULATION ON TIMING OF ARCTIC DINOSAUR NESTING

The timing of the birth of mammals and birds in the Arctic occurs in the spring around the onset of food availability for young (Figure 7.16). For one of the important mammalian vertebrates of the Arctic, the caribou, the timing of calving is typically some weeks before the advent of the greening of the landscape (Bergerud et al.,

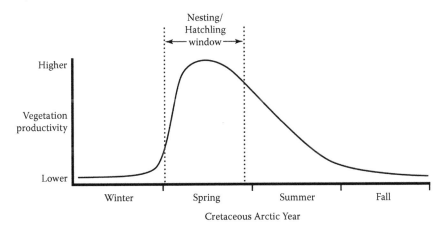

**FIGURE 7.16**   Timing of the nesting and hatchling window in a year in the Cretaceous Arctic.

2007). In contrast, for Arctic birds hatching can be coincident with the onset of the greening of the landscape (Van der Graaf et al., 2006; Van der Jeugd et al., 2009) or, in the case of insectivorous birds, just prior to the maximum availability of arthropods (Tulp and Schekkerman, 2008).

Today, the marked seasonality that characterizes the Arctic affects patterns of nesting among birds that live year round within the ecosystem as well as migratory birds (Pielou, 1994). Seasonal patterns of resource availability also regulated timing patterns of nesting among resident dinosaurs of the Cretaceous Arctic of Alaska.

The end members for Arctic avian breeders are either altricial or precocial species (Pielou, 1994). Altricial species are birds born effectively helpless; they are naked, blind, and incapable of walking. There is therefore a significant window in which altricial birds develop after hatching while still in their nest. They must be fed and cared for. In precocial species, babies develop longer in their eggs before hatching; thus they are able to see, walk or run shortly after hatching, and are better able to thermoregulate. Non-avian dinosaurian growth patterns did not approach those observed among modern altricial birds; rather, they show similarity to precocial birds (Erickson et al., 2001), and histological data suggest that altricial dinosaurs were rare (Horner, 2000).

As discussed earlier, a growing body of biomechanical, isotopic, and histological evidence suggests that Arctic dinosaurs did not have long seasonal migrations (Chapter 4). Therefore, following the patterns of modern birds, one can speculate the patterns of timing of nesting might follow that of precocial species. In Montana, at the famous dinosaur locality known as Egg Mountain, *Troodon* nests have been discovered showing that this dinosaur nested on the ground (Varricchio et al., 1997) and that nests appear to have been unprotected, as in precocial rather than altricial species.

In the Arctic, approximately 50% of bird species are precocial (Pielou, 1994). At first glance that would seem not to be very helpful in speculating about Cretaceous Arctic dinosaurs. However, migratory Arctic birds tend to be altricial, while residents tend to be precocial (Pielou, 1994). The altricial birds of the Arctic nest in well-built, well-hidden protected places such as rocky ledges, precipices, crevices, or at the edges of marshes with thick vegetation. Perhaps it is because altricial young generally cannot thermoregulate as well as precocial young (Elphick et al., 2001) that Arctic altricial birds start their breeding later than precocial species (Pielou, 1994).

Subtracting those modern Arctic altricial birds that are not ground nesters, we can then interpret that the majority of the remaining Arctic birds are precocial. If *Troodon* was precocial, and the longer incubation period of Arctic precocial birds was also a generality including *Troodon*, then given the profound seasonality of the ancient Arctic, *Troodon* likely had to begin nesting early, perhaps just as the ancient Arctic Spring was beginning. Further, because of the phylogenetic position of birds as the living dinosaurs, that nesting and hatching likely coincided with the initiation of the greening of the landscape due to the seasonal growth of plants. This cyclical pattern was likely true regardless of whether the breeding dinosaurs ate those plants or fed on herbivorous dinosaurs.

## SUMMARY

The discovery of abundant dinosaur bones and footprints across the Alaskan high latitudes prompts many questions about the paleobiology of these animals. Perhaps foremost is the issue of whether they engaged in seasonal migration to avoid the harshest rigors of the Arctic winter. A growing body of evidence indicates that dinosaurs were year-round residents of the ancient north. There is ample evidence from the fossil bone record as well as the fossil footprint record that the herbivorous dinosaurs were gregarious and sustained large multigenerational herds.

Evidence that Alaskan dinosaurs partitioned food resources suggests ecological complexity. However, we still do not fully understand the dietary differences between these northern animals and their counterparts in the lower latitudes, as indicated by tooth microwear.

The issue of the color of Arctic dinosaurs is also still unresolved. Coloration of organisms often has an important role in predation, communication, and metabolism. However, despite the general awareness of Gloger's Rule—that animals living in lower latitudes tend to be more darkly colored than those living in higher latitudes—there is no consensus as to why some modern animals living in the Arctic seasonally turn white. Regardless of the reasons modern animals turn white in the winter, the temperature range in which some of them do change color is comparable to the reconstructed temperatures for the ancient Arctic. Given that low temperatures in those modern environments inhibit the production of melanin, it is entirely reasonable to assume that some of the Cretaceous Arctic dinosaurs also changed to white in the winter.

The timing of the birth of mammals and birds in the Arctic occurs in the spring at the onset of the time food is readily available for young. Similarly, the beginning of the arctic dinosaur breeding season, regardless of herbivorous or carnivorous diet, likely coincided with the initiation of the greening of the landscape during the spring.

The modern Arctic is an ecosystem dominated by profound seasonal fluctuations in solar energy, temperature, and biological productivity. How dinosaurs survived and the dynamics between these organisms and their environment are driving issues in understanding an ancient arctic ecosystem. While some aspects of Arctic dinosaur paleobiology are beginning to be understood, exactly how arctic dinosaurs survived a full polar year remains the big unresolved question. That question is what makes this field so fascinating, as it challenges everything we think we know about the paleobiology of dinosaurs.

## REFERENCES

Bell, P.R. and E. Snively. 2008. Polar dinosaurs on parade: A review of dinosaur migration. *Alcheringa* 32:271–284.

Bergerud, A.T., S.N. Luttich, and L. Camps. 2007. *The Return of Caribou to Ungava.* Montreal: McGill-Queen's University Press.

Blix, A.S. 2005. *Arctic Animals and Their Adaptations to Life on the Edge.* Trondheim: Tapir Academic Press.

Brinkman, D.B., M.J. Ryan, and D.A. Eberth. 1998. The paleogeographic and stratigraphic distribution of ceratopsids (Ornithischia) in the upper Judith River Group of western Canada. *Palaios* 13:160–169.

Brouwers, E.M., W.A. Clemens, R.A. Spicer, T.A. Ager, L.D. Carter, and W.V. Sliter. 1987. Dinosaurs on the North Slope, Alaska: Reconstructions of high-latitude, latest Cretaceous environments. *Science* 237:1608–1610.

Buckland, W. 1829. On the discovery of a new species of Pterodactyle; and also of the Faeces of the Ichthyosaurus; and of a black substance resembling Sepia, or Indian Ink, in the Lias at Lyme Regis. *Proceedings of the Geological Society of London* 1:96–98.

Chin, K. 2002. Analyses of coprolites produced by carnivorous vertebrates. *Paleontological Society Papers* 8:43–50.

Chin, K. 2007. The paleobiological implications of herbivorous dinosaur coprolites from the Upper Cretaceous Two Medicine Formation of Montana: Why eat wood? *Palaios* 22:554–566.

Chin, K. and B.D. Gill. 1996. Dinosaurs, dung beetles, and conifers: Participants in a Cretaceous food web. *Palaios* 11:280–285.

Chin, K., T.T. Tokaryk, G.M. Erickson, and L.C. Calk. 1998. A king-sized theropod coprolite. *Nature* 393:680–682.

Chinsamy, A., D.B. Thomas, A.R. Tumarkin-Deratzian, and A.R. Fiorillo. 2012. Hadrosaurs were perennial polar residents. *The Anatomical Record* 295:610–614.

Clark, J.M., T. Maryanska, and R. Barsbold. 2004. Therizinosauroidea. In *The Dinosauria*, 2nd edition, eds. D.B. Weishampel, P. Dodson, and H. Osmolska, 151–164. Berkeley: University of California Press, Berkeley.

Clemens, W.A. and L.G. Nelms. 1993. Paleoecological implications of Alaskan terrestrial vertebrate fauna in latest Cretaceous time at high latitudes. *Geology* 21:503–506.

Currie, P.J. 1989. Long-distance dinosaurs. *Natural History* 6:59–65.

Currie, P.J. 1998. Possible evidence of gregarious behavior in tyrannosaurids. *GAIA* 15:271–277.

Currie, P.J. and D.A. Eberth. 2010. On gregarious behavior in *Albertosaurus*. *Canadian Journal of Earth Sciences* 47:1277–1289.

Cuthbertson, R.S., A. Tirabasso, N. Rybczynski, and R.B. Holmes. 2012. Kinetic limitations of intracranial joints in *Brachylophosaurus canadensis* and *Edmontosaurus regalis* (Dinosauria: Hadrosauridae), and their implications for the chewing mechanics of hadrosaurids. *The Anatomical Record* 295:968–979.

Dekay, J.E. 1830. On the discovery of coprolites in North America. *Philosophical Magazine* 7:321–322.

Duffin, C.J. 2012. The earliest published records of coprolites. *New Mexico Museum of Natural History and Science Bulletin* 57:25–28.

Eberth, D.A. and P.J. Currie. 2010. Stratigraphy, sedimentology, and taphonomy of the Albertosaurus bonebed (upper Horseshoe Canyon Formation; Maastrichtian), southern Alberta, Canada. *Canadian Journal of Earth Sciences* 47:1119–1143.

El-Baz, F. 1968. Coprolites vs. fecal pellets. *American Association of Petroleum Geologists Bulletin* 52:526.

Elphick, C., J.B. Dunning, Jr., and D.A. Sibley. 2001. *The Sibley Guide to Bird Life & Behavior*. New York: Alfred A. Knopf, Inc.

Erickson, G.M. and P.S. Druckenmiller. 2011. Longevity and growth rate estimates for a polar dinosaur: A *Pachyrhinosaurus* (Dinosauria: Neoceratopsia) specimen from the North Slope of Alaska showing a complete developmental record. *Historical Biology* 23:327–334.

Erickson, G.M., K.C. Rogers, and S.A. Yerby. 2001. Dinosaurian growth patterns and rapid avian growth rates. *Nature* 412:429–433.

Fiorillo, A.R. 1998. Dental micro wear patterns of the sauropod dinosaurs camarasaurus and diplodocus: Evidence for resource partitioning in the late Jurassic of North America. *Historical Biology* 13(1):1–16.

Fiorillo, A.R. 2008. On the occurrence of exceptionally large teeth of *Troodon* (Dinosauria: Saurischia) from the Late Cretaceous of northern Alaska. *Palaios* 23:322–328.

Fiorillo, A.R. 2011. Microwear patterns on the teeth of northern high latitude hadrosaurs with comments on microwear patterns in hadrosaurs as a function of latitude and seasonal ecological constraints. *Palaeontologia Electronica* 14(3):20A, 17p. palaeo-electronica. org/2011_3/7_fiorillo/index.html

Fiorillo, A.R. and T.L. Adams. 2012. A therizinosaur track from the Lower Cantwell Formation (Upper Cretaceous) of Denali National Park, Alaska. *Palaios* 27:395–400.

Fiorillo, A.R., M. Contessi, Y. Kobayashi, and P.J. McCarthy. 2014b. Theropod tracks from the Lower Cantwell Formation (Upper Cretaceous) of Denali National Park, Alaska, USA with comments on theropod diversity in an ancient, high-latitude terrestrial ecosystem. *New Mexico Museum of Natural History and Science Bulletin* 62:429–439.

Fiorillo, A.R. and R.A. Gangloff. 2000. Theropod teeth from the Prince Creek Formation (Cretaceous) of northern Alaska, with speculations on arctic dinosaur paleoecology. *Journal of Vertebrate Paleontology* 20:675–682.

Fiorillo, A.R. and R.A. Gangloff. 2001. The caribou migration model for Arctic hadrosaurs (Ornithischia: Dinosauria): A reassessment. *Historical Biology* 15:323–334.

Fiorillo, A.R., S.T. Hasiotis, and Y. Kobayashi. 2014a. Herd structure in Late Cretaceous polar dinosaurs: A remarkable new dinosaur tracksite, Denali National Park, Alaska, USA. *Geology* 42:719–722.

Fiorillo, A.R., P.J. McCarthy, and P.P. Flaig. 2010b. Taphonomic and sedimentologic interpretations of the dinosaur-bearing Upper Cretaceous Strata of the Prince Creek Formation, Northern Alaska: Insights from an ancient high-latitude terrestrial ecosystem. *Palaeogeography, Palaeoclimatology, Palaeoecology* 295:376–388.

Fiorillo, A.R., P.J. McCarthy, and P.P. Flaig. 2016. A multi-disciplinary perspective on habitat preferences among dinosaurs in a Cretaceous Arctic greenhouse world, North Slope, Alaska (Prince Creek Formation: Lower Maastrichtian). *Palaeogeography, Palaeoclimatology, Palaeoecology* 441:377–389.

Fiorillo, A.R., P.J. McCarthy, P.P. Flaig, E. Brandlen, D.W. Norton, P. Zippi, L. Jacobs, and R.A. Gangloff. 2010a. Paleontology and paleoenvironmental interpretation of the Kikak-Tegoseak Quarry (Prince Creek Formation: Late Cretaceous), northern Alaska: A multi-disciplinary study of a high-latitude ceratopsian dinosaur bonebed. In *New Perspectives on Horned Dinosaurs*, eds. M.J. Ryan, B.J. Chinnery-Allgeier, and D.A. Eberth, 456–477. Bloomington: Indiana University Press.

Fiorillo, A.R. and R.S. Tykoski. 2012. A new species of centrosaurine ceratopsid *Pachyrhinosaurus* from the North Slope (Prince Creek Formation: Maastrichtian) of Alaska. *Acta Palaeontologica Polonica* 57:561–573.

Fiorillo, A.R. and R.S. Tykoski. 2013. An immature *Pachyrhinosaurus perotorum* (Dinosauria: Ceratopsidae) nasal reveals unexpected complexity of craniofacial ontogeny and integument of *Pachyrhinosaurus*. *PLoS ONE* 8(6):e65802. doi:10.1371/journal.pone.0065802.

Fiorillo, A.R. and R.S. Tykoski. 2014. A diminutive new tyrannosaur from the top of the world. *PLoS ONE* 9(3):e91287. doi:10.1371/journal.pone.0091287.

Flaig, P.P., P.J. McCarthy, and A.R. Fiorillo. 2013. Anatomy, evolution and paleoenvironmental interpretation of an ancient Arctic coastal plain: Integrated paleopedology and palynology from the Upper Cretaceous (Maastrichtian) Prince Creek Formation, North Slope, Alaska, USA. In *New Frontiers in Paleopedology and Terrestrial Paleoclimatology: Paleosols and Soil Surface Analogue Systems*, eds. S.G. Driese and L.C. Nordt, 179–230. Tulsa OK: SEPM Special Publication 104.

Fowler, D.W., E.A. Freedman, J.B. Scannella, and R.E. Kambic. 2011. The predatory ecology of *Deinonychus* and the origin of flapping in birds. *PLoS ONE* 6(12):e28964.

Fricke, H.C., R.R. Rogers, and T.A. Gates. 2009. Hadrosaurid migration: Inferences based on stable isotope comparisons among Late Cretaceous dinosaur localities. *Paleobiology* 35:270–288.

Gangloff, R.A. 2012. *Dinosaurs Under the Aurora*. Bloomington, IN: Indiana University Press.

Gangloff, R.A. and A.R. Fiorillo. 2010. Taphonomy and paleoecology of a bonebed from the Prince Creek Formation, North Slope, Alaska. *Palaios* 25(5):299–317.

Geisel, T.G. ["Dr. Seuss"]. 1955. *On Beyond Zebra!* New York: Random House.

Goswami, A., J.J. Flynn, L. Ranivoharimanana, and A.R. Wyss. 2005. Dental microwear in Triassic amniotes: Implications for paleoecology and masticatory mechanics. *Journal of Vertebrate Paleontology* 25:320–329.

Grinnell, J. 1917. The niche-relationships of the California Thrasher. *Auk* 34:427–433.

Hamilton, W.J., III. 1973. *Life's Color Code.* New York: McGraw-Hill Book Company.

Hamilton, W.J., III, and F. Heppner. 1967. Radiant solar energy and the function of black homeotherm pigmentation: An hypothesis. *Science* 155:196–197.

Hodge, R.P. 1976. *Amphibians & Reptiles in Alaska, The Yukon & Northwest Territories.* Anchorage: Alaska Northwest Publishing Company.

Hollocher, K.T., O.A. Alcober, C.E. Colombi, and C.T. Hollocher. 2005. Carnivore coprolites from the Upper Triassic Ischigualasto Formation, Argentina: Chemistry, mineralogy, and evidence for rapid initial mineralization. *Palaios* 20:51–63.

Horner, J.R. 2000. Dinosaur reproduction and parenting. *Annual Review of Earth and Planetary Science* 28:19–45.

Hotton, N., III. 1980. An alternative to dinosaur endothermy. In *A Cold Look at Warm-Blooded Dinosaurs*, eds. R.D.K. Thomas and E.C. Olson, 311–350. Boulder, CO: Westview Press.

Hunt, A.P. 1992. Late Pennsylvanian coprolites from the Kinney Brick Quarry, central New Mexico, with notes on the classification and utility of coprolites. *New Mexico Bureau of Mines and Mineral Resources Bulletin* 138:221–229.

Hunt, A.P., K. Chin, and M.G. Lockley. 1994. The palobiology of vertebrate coprolites. In *The Palaeobiology of Trace Fossils*, ed. S. Donovan, 221–240. London: John Wiley.

Jain, S.L. 1983. Spirally coiled "coprolites" from the Upper Triassic Maleri formation, India. *Palaeontology* 26:813–829.

King, C.M. and J.E. Moody. 1982. The biology of the stoat (*Mustela erminea*) in the National Parks of New Zealand. V. Moult and colour change. *New Zealand Journal of Zoology* 9:119–130.

Li, Q., K.-Q. Gao, J. Vinther, M.D. Shawkey, J.A. Clarke, L. D'Alba, Q. Meng, D.E.G. Briggs, and R.O. Prum. 2010. Plumage color patterns of an extinct dinosaur. *Science* 327:1369–1372.

Li, R., M.G. Lockley, P.J. Makovicky, M. Matsukawa, M.A. Norell, J.D. Harris, and M. Liu. 2008. Behavioral and faunal implications of Early Cretaceous deinonychosaur trackways from China. *Naturwissenschaften* 95:185–191.

Lister, M. 1678. *Historiae Animalium Angliae Tres Tractatus. Unus de Araneis. Alter de Cochleis tum Terrestribus tum Fluviatilibus. Tertius de Cochleis Marinis.* London: Martyn.

Ludvigson, G., L.A. González, R.A. Metzger, B.J. Witzke, R.L. Brenner, A.P. Murillo, and T.S. White. 1998. Meteoric sphaerosiderite lines and their use for paleohydrology and paleoclimatology. *Geology* 26:1039–1042.

Manitoba Agriculture, Food and Rural Development. 2015. Properties of manure. *Manitoba Agriculture, Food and Rural Development* 1–42.

Manning, P.L., L. Margetts, M.R. Johnson, P.J. Withers, W.I. Sellers, P.L. Falkingham, P.M. Mummery, P.M. Barrett, and D.R. Raymont. 2009. Biomechanics of dromaeosaurid dinosaur claws: Application of x-ray microtomography, nanoindentation, and finite element analysis. *The Anatomical Record* 292:1397–1405.

Manning, P.L., D. Payne, J. Pennicott, P.M. Barrett, and R.A. Ennos. 2006. Dinosaur killer claws or climbing crampons? *Biology Letters* 2:110–112.

Mantell, G.A. 1822. *The Fossils of the South Downs; or Illustrations of the Geology of Sussex.* London: Lupton Relfe.

Marchand, P.J. 1996. *Life in the Cold: An Introduction to Winter Ecology*, 3rd edition. Hanover, NH: University Press of New England.

Maxwell, W.D. and J.H. Ostrom. 1995. Taphonomy and paleobiological implications of *Tenontosaurus–Deinonychus* associations. *Journal of Vertebrate Paleontology* 15:707–712.

McAllister, J.A. 1985. Reevaluation of the formation of spiral coprolites. *University of Kansas Paleontological Contributions* 114:1–12.

McKay, M.E.P. 2008. Paleoecologies of the mammalian fossils from fossil faunas of Natural Trap Cave and Little Box Elder Cave, Wyoming. *MS thesis*, University of South Carolina.

Nichols, L. and F.L. Bunnell. 1999. Natural history of thinhorn sheep. In *Mountain Sheep of North America*, eds. R. Valdez and P.R. Krausman, 23–77. Tucson, AZ: University of Arizona Press.

Northwood, C. 2005. Early Triassic coprolites from Australia and their palaeobiological significance. *Palaeontology* 48:49–68.

Novacek, M.J. 1996. Paleontological data and the study of adaptation. In *Adaptation*, eds. M.R. Rose and G.V. Lauder, 311–359. San Diego, CA: Academic Press.

Ostrom, J.H. 1969. Osteology of *Deinonychus antirrhopus*, an unusual theropod from the Lower Cretaceous of Montana. *Bulletin of the Peabody Museum* 30:1–165.

Parrish, J.M., J.T. Parrish, J.H. Hutchison, and R.A. Spicer. 1987. Late Cretaceous vertebrate fossils from the North Slope of Alaska and implications for dinosaur ecology. *Palaios* 2:377–389.

Paul, G.S. 1988. Physiological, migratorial, climatological, geophysical, survival, and evolutionary implications of Cretaceous polar dinosaurs. *Journal of Paleontology* 62:640–652.

Pettygrove, G.S., A.L. Heinrich, and A.J. Eagle. 2010. Dairy manure nutrient content and forms. *University of California Cooperative Extension Manure Technical Bulletin Series* 1–10.

Pielou, E.C. 1994. *A Naturalist Guide to the Arctic*. Chicago, IL: University of Chicago Press.

Pitelka, F.A. 1966. Some characteristics of microtine cycles in the Arctic. In *Arctic Biology*, ed. H.P. Hansen, 153–184. Corvallis, OR: Oregon State University Press, Corvallis.

Rappole, J.H. 2013. *The Avian Migrant: The Biology of Bird Migration*. New York, NY: Columbia University Press.

Repenning, C. 2004. Email correspondence to Fiorillo, 11/12/2004.

Ricklefs, R.E. and G.L. Miller. 2000. *Ecology*, 4th edition. New York, NY: W.H. Freeman and Company.

Roach, B.T. and D.B. Brinkman. 2007. A reevaluation of cooperative pack hunting and gregariousness in *Deinonychus antirrhopus* and other nonavian theropod dinosaurs. *Bulletin of the Peabody Museum* 48:103–138.

Rogers, R.R., D.A. Eberth, and A.R. Fiorillo (eds.). 2007. *Bonebeds: Genesis, Analysis, and Paleobiological Significance*. Chicago: University of Chicago Press.

Rouse, G.E. and S.K. Svrivastava. 1972. Palynological zonation of cretaceous and early tertiary rocks of the bonnet plume formation northeastern Yukon, Canada. *Canadian Journal of Earth Sciences* 9:1163–1179.

Savile, D.B.O. 1972. Arctic adaptations in plants. *Canada Department of Agriculture, Research Branch, Monograph* 6:1–81.

Scholander, P.F., V. Walters, R. Hock, and L. Irving. 1950. Adaptation to cold in arctic and tropical mammals and birds in relation to body temperature, insulation, and basal metabolic rate. *The Biological Bulletin* 99:259–271.

Sereno, P.C. 1998. A rationale for phylogenetic definitions, with application to the higher-level taxonomy of Dinosauria. *Neues Jahrbuch für Geologie und Paläontologie Abhandlungen* 210:41–83.

Seton, E.T. 1911. *The Arctic Prairies: A Canoe Journey of 2,000 Miles in Search of the Caribou: Being the Account of a Voyage to the Region North of Aylmer Lake*. Toronto: Briggs.

Sibley, D.A. 2000. *The Sibley Guide to Birds*. New York, NY: Alfred A. Knopf, Inc.

Suarez, C.A., G.A. Ludvigson, L.A. González, A.R. Fiorillo, P.P. Flaig, and P.J. McCarthy. 2013. Use of multiple oxygen isotope proxies for elucidating Arctic cretaceous palaeohydrology. In *Isotopic Studies in Cretaceous Research*, eds. A.-V. Bojar, M.C. Melinte-Dobrinescu, and J. Smit, 185–202. Geological Society, London, Special Publications 382.

Suarez, M.B., L.A. González, and G.A. Ludvigson. 2011. Quantification of greenhouse hydrologic cycle from equatorial to polar latitudes: The mid-Cretaceous water bearer revisited. *Palaeogeography, Palaeoclimatology, Palaeoecology* 307:302–312.

Thulborn, R.A. 1991. Morphology, preservation and palaeobiological significance of dinosaur coprolites. *Palaeogeography, Palaeoclimatology, Palaeoecology* 83:341–366.

Tulp, I. and H. Schekkerman. 2008. Has prey availability for arctic birds advanced with climate change? Hindcasting the abundance of tundra arthropods using weather and seasonal variation. *Arctic* 61:48–60.

Ufnar, D.F., L.A. González, G.A. Ludvigson, R.L. Brenner, and B.J. Witzke. 2004. Evidence for increased latent heat transport during the Cretaceous (Albian) greenhouse warming. *Geology* 32:1049–1052.

Valdez, R. and P.R. Krausman. 1999. Description, distribution, and abundance of mountain sheep in North America. In *Mountain Sheep of North America*, eds. R. Valdez and P.R. Krausman, 3–22. Tucson, AZ: University of Arizona Press.

Van der Graaf, A.J., J. Stahl, A. Klimkowska, J.P. Bakker, and R.H. Drent. 2006. Surfing on a green wave—How plant growth drives spring migration in the Barnacle Goose *Branta leucopsis. Ardea* 94:567–577.

Van der Jeugd, H.P., G. Eichorn, K.E. Litvins, J. Stahl, K. Larsson, A.J. Van der Graaf, and R.H. Drent. 2009. Keeping up with early springs: Rapid range expansion in an avian herbivore incurs a mismatch between reproductive timing and food supply. *Global Change Biology* 15:1057–1071.

Van Valkenburgh, B. and R.E. Molnar. 2002. Dinosaurian and mammalian predators compared. *Paleobiology* 28:527–543.

Varricchio D.J., F. Jackson, J.J. Borkowsk, and J.R. Horner. 1997. Nest and egg clutches of the dinosaur *Troodon formosus* and the evolution of avian reproductive traits. *Nature* 385:247–250.

Vibe, C. 1967. Arctic animals in relation to climatic fluctuations. The Danish Zoogeographical Investigations in Greenland. *Meddelelser om Grønland* 170:1–227.

Vinther, J., D.E.G. Briggs, J. Clarke, G. Mayr, and R.O. Prum. 2010. Structural coloration in a fossil feather. *Biology Letters* 6:128–131.

Ward, J.M., G.D. Ruxton, D.C. Houston, and D.J. McCafferty. 2007. Thermal consequences of turning white in winter: A comparative study of red grouse *Lagopus lagopus scoticus* and Scandinavian willow grouse *L. l. lagopus. Wildlife Biology* 13:120–129.

Weishampel, D.B., P.M. Barrett, R.A. Coria, J. Le Loeuff, X. Xu, X. Zhoa, A. Sahni, E.M.P. Gomani, and C.R. Noto. 2004. Dinosaur distribution, 517–606. In *The Dinosauria*, 2nd edition, eds. H. Osmolska, 517–606. Berkeley, CA: University of California Press.

Western, D., C. Moss, and N. Georgiadis. 1983. Age estimation and population age structure of elephants from footprint dimensions. *Journal of Wildlife Management* 47:1192–1197.

Williams, M.E. 1972. The origin of "spiral coprolites." *The University of Kansas Paleontological Contributions* 59:1–19.

Zelenitsky, D.K., F. Therrien, and Y. Kobayashi. 2009. Olfactory acuity in theropods: Palaeobiological and evolutionary implications. *Proceedings of the Royal Society B* 276:267–273.

Zhang, F., S.L. Kearns, P.J. Orr, M.J. Benton, Z. Zhou, D. Johnson, X. Xu, and X. Wang. 2010. Fossilized melanosomes and the colour of Cretaceous dinosaurs and birds. *Nature* 463:1075–1078.

# 8 Paleoclimate

## INTRODUCTION

The role of paleontology in framing the original questions about changes in climate is worth reviewing. Reconstructions of past climates have a venerable history. The 11th century (Common Era) Chinese polymath, Shen Kuo, determined that plant fossils in east-central China were the remains of bamboo and proposed that since no bamboo was found in this region at the time of his analysis, the climate was likely to have changed (Chaloner and Crebin, 1990). Even though his taxonomy was incorrect, Shen Kuo's conclusion was remarkable given the scientific framework of his day.

Later, the discovery of fossil elephants in Siberia and at Big Bone Lick, Kentucky, in the mid-1700s challenged naturalists of the day by opening questions about past climates (Guthrie, 1990; Hedeen, 2008; Thomson, 2008). These animals resemble modern elephants living in Africa's and southern central Asia's warmer climates, but clearly these fossil forms were discovered in areas much colder today. To explain these fossils in North America, the American Benjamin Franklin (1767) wrote a letter to George Croghan in which he articulated that climates change by questioning if "the earth had anciently been in another position, and the climates differently placed from what they are at present."

Among the primary questions presented by these fossil elephants was how they got buried in these colder environments. The essence of the debate regarding the origin of fossil mammoths in northeastern Asia is, as Guthrie (1990) expressed it, that the carcasses represented either "Floaters or Sinkers." Whereas some naturalists advocated a flotation model whereby these elephants lived in the south and floated to their burial sites in the north, Cuvier (1825) proposed that these animals were in fact residents of the north and that they were adapted to polar climates. In other words, global climate had changed over time.

Frederick Beechey wrote of the co-occurrence of the bones of elephants, musk-oxen, and reindeer along the shores of Eschscholtz Bay in Alaska (Beechey, 1831), inferring something about the climate of the time that could accommodate these three animals. On the other hand, William Buckland (1831) pointed out that these bones were found between the high and low water levels and that the bones were likely a mixed assemblage of remains. This discussion was an attempt to use fossil vertebrates to gain insight into past climates.

Indeed, for over 100 years, not only have fossils been used for identifying climate change, but it has been suggested that oscillations in climate, and specifically the role of aridity, have shaped the pattern of evolution (Matthew, 1915; Barrell, 1916). The central point of this discussion is that fossils—especially fossil vertebrates—had, and continue to have, a role in understanding the nature of past climate régimes.

Organisms themselves may have a major regulatory control on Earth's climate. Indeed, a casual survey of science stories in popular media today would reveal the most prevalent discussion topic concerns what role humans have played in altering climate, and the recognition that humans must alter their behavior as a species to accommodate the climate change they are bringing about. As a broad concept, this is an example that ecosystem effects on climate create feedback loops. More specifically, biological processes play a major role in the carbon cycle—the global exchange of carbon between the biosphere, the lithosphere, the hydrosphere, and the atmosphere (Figure 8.1). Shifts in the balance of carbon through this cycle can alter weather patterns and oceanic and atmospheric chemistry, and result in changes in climate. In monitoring the carbon balances within the carbon cycle, climate studies often focus on parameters such as methane input, fire frequency, temperature reconstruction, moisture regimes, and carbon dioxide levels to understand past patterns in order to understand future directions of climate.

As paleontologists and climatologists can attest, climates have changed, sometimes radically, over the course of geologic time. Climate systems result from geophysical and geochemical feedbacks among atmospheric composition, oceans, topography, vegetation, cloud cover, moisture, and biomass (Figure 8.2). The driving question for ancient climate change is how to understand these variables in deep time.

Proxies are commonly used for understanding past climates, and the use of proxies can vary. Though temperature as shorthand for climate has fallen under some recent criticism (e.g., Evans et al., 2005; Clarke and Gaston, 2006), reconstructing temperature regimes still dominates thinking and general circulation models of modern and ancient climates. Much of the attention centered on climate and climate change is directed toward predicting and responding to future climate changes.

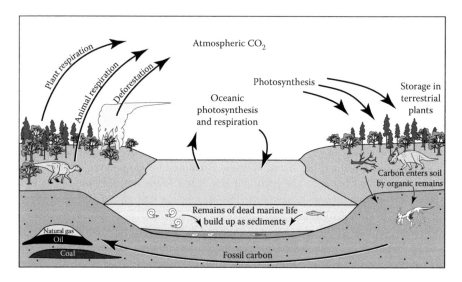

**FIGURE 8.1**  A representation of the carbon cycle showing how carbon can be passed among the biosphere, the atmosphere, and the lithosphere.

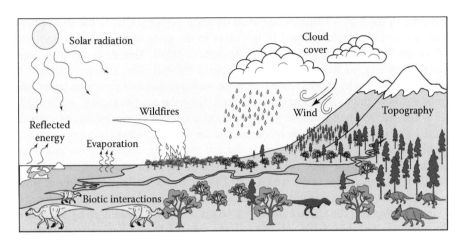

**FIGURE 8.2** Abiotic and biotic components of the climate system. Earth climate studies focus on a wide array of interactions and feedback between these various components.

The Cretaceous pole-to-equator temperature gradient was reduced from that observed today (e.g., DeConto et al., 1999), and there are very clearly variations through geologic time in warming and cooling trends that occurred within the Cretaceous even within the high latitudes (e.g., Salazar-Jaramillo et al., 2016). Precipitation gradients from pole to equator during the Cretaceous were steeper than now (Ludvigson et al., 1998; Ufnar et al., 2002; Suarez et al., 2011). Under warm "greenhouse" conditions in Earth history, the hydrologic cycle was intensified (Barron et al., 1989). So, given this close relationship between the hydrologic cycle and climate, a great deal of effort focused on reconstructing past Cretaceous climate has incorporated geochemical proxies in understanding ancient high-latitude climate conditions (e.g., Ufnar et al., 2001, 2002, 2004; Suarez et al., 2011, 2013, 2016; Salazar-Jaramillo et al., 2015, 2016). For example, iron-carbonate nodules formed in ancient soils, or pedogenic siderites, provide details of the water derived through ancient precipitation, and such siderites have been sampled throughout the Western Interior Basin of North America in an effort to better understand Cretaceous paleoclimate and climate modeling (e.g., Ludvigson et al., 1998; Ufnar et al., 2001, 2004; Suarez et al., 2011).

A water molecule ($H_2O$) with oxygen isotope 18 is heavier than the normal water molecule containing oxygen isotope 16. As such, in clouds, the molecules with heavier oxygen isotope tend to condense from vapor to liquid before the lighter molecules, and the rains are isotopically enriched with the heavy isotope. Then, processes such as photosynthesis or the formation of soil or other surface minerals such as siderite that require water reflect the isotopic composition and heavy isotope enrichment of the rain.

With respect to the Cretaceous Arctic, a comparison of the stable oxygen isotope signal from siderites to that found in dinosaur tooth enamel phosphate from the Prince Creek Formation suggests that heavy oxygen-depleted precipitation (oxygen-18) occurred at these ancient high latitudes during the deposition of this rock unit (Suarez et al., 2013). Manabe and Wetherald (1980) modeled an increase in moisture in the higher latitudes due to an increase in global $CO_2$ in the atmosphere. This moisture increase resulted in increases in the rates of precipitation and run-off in the high latitudes (Manabe and Wetherald, 1980). This depletion in isotopically heavy oxygen was attributed to increased rainfall from an intensification of the hydrologic cycle, and that intensification was connected to enhanced latent heat transport from the warmer southern regions to the polar regions (Suarez et al., 2013).

While studies such as those mentioned above tie into global models of climate, more locally, humid conditions reconstructed for northern Alaska were variable and highly seasonal. This interpretation is derived from data obtained from paleosols, ancient soils that have since been buried and turned to rock. Paleosols, like modern soils, record the climate under which they formed. In one of the few studies of fossil soils formed within an ancient Arctic ecosystem, Flaig and others (2013) were able to demonstrate that the Cretaceous Arctic coastal plain was influenced by seasonally fluctuating water tables and floods. Alternating wet and dry seasonal cycles influenced the highly unusual formation of the clay mineral illite (Salazar Jaramillo et al., 2016), normally formed through diagenesis, those physical and chemical processes responsible for transforming sediment into rock. Further, in contrast to modern polar environments, there was no evidence of cryogenic, or ice-related, soil-forming processes in this ancient warm Arctic setting (Spicer and Parrish, 1990; Flaig et al., 2013).

Paleontologically based climate proxies such as poleward extensions in geographic range of temperate flora and invertebrates indicate that the Cretaceous northern high latitudes experienced extreme warmth compared to modern times, and there was a reduced equator-to-pole temperature gradient (e.g., Wolfe and Upchurch, 1987; Spicer and Parrish, 1990; Spicer, 2003; Jenkyns et al., 2004; Tomsich et al., 2010).

An example of poleward extension of geographic range is the remarkable occurrence of invertebrate trace fossils of Cretaceous mole crickets from the Lower Cantwell Formation in Denali National Park (Fiorillo et al., 2009, 2014a; Figure 8.3). Mole crickets are insects related to grasshoppers, locusts, and crickets. Because of their negative agricultural impacts, these are a well-studied group of animals. Modern mole crickets are found on all continents except Antarctica, and are found only in lower and mid-latitudes. The occurrence of their fossil traces in the Lower Cantwell Formation is further evidence that the climate was warmer during the Cretaceous than the climate today (Fiorillo et al., 2014a).

Paleobotanical contributions to our understanding of much warmer Cretaceous Arctic temperatures come from the study of the relationship between leaf margins and temperature (e.g., Bailey and Sinnott, 1915, 1916; Wolfe, 1979). The basic point of this relationship is that within a population of deciduous plant species with varying shapes of leaves, the relative abundance of smooth-margined leaves compared to tooth-margined, or jagged-edged, leaves is a function of temperature. The correlation is strikingly strong. Simply put, the higher the percentage of smooth-margined leaves, the warmer the climate (Figure 8.4). Why this pattern occurs in nature remains poorly understood but may have to do with the fact that plants in the higher latitudes have a much shorter growing season, therefore they need to begin photosynthesis quickly once the leaves grow. A jagged edge to a tooth enables more water to move out of the leaf, which in turn would increase the flow of sap, which would increase the rate of photosynthesis.

**FIGURE 8.3** Cretaceous mole cricket burrow from the Lower Cantwell Formation of Denali National Park, Alaska. (Photograph courtesy of Stephen Hasiotis, University of Kansas.)

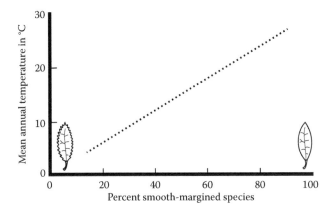

**FIGURE 8.4** Graph showing the relationship between leaf margin morphology and mean annual temperature.

Understanding that such a pattern exists, this powerful tool was first applied to Cretaceous Arctic floras by Spicer and Parrish (1986, 1990) to estimate mean annual paleotemperature. Their estimate for mean annual temperature in northern Alaska during the Campanian to Maastrichtian was about 5°C, a temperature reconstruction that has been slightly modified to 6–7°C with subsequent work (Herman et al., 2016). A comparable temperature of approximately 7°C, based on the fossil flora from the Lower Cantwell Formation of Denali National Park, has also been reconstructed (Tomsich et al., 2010) which shows a somewhat equable temperature regime across Alaska during the Cretaceous. For perspective, the mean annual temperature in Seattle, Washington, is 11°C, and that in Calgary, Alberta, is 4°C, so these paleotemperatures indicate a more moderate temperature during the Cretaceous for Alaska. To account for these warmer Cretaceous temperatures, numerous mechanisms have been explored, including increased ocean heat transport from the lower latitudes to the higher latitudes (Herman and Spicer, 1996; Johnson et al., 1996), the evolution of continental paleogeography (Donnadieu et al., 2006), increased atmospheric greenhouse gases (Bice et al., 2006), the physical properties of clouds (Kump and Pollard, 2008), and the role of terrestrial vegetation (Otto-Bliesner and Upchurch, 1997; DeConto et al., 2000; Zhou et al., 2012). None of these proposed mechanisms has been wholly embraced because either their climate impact was insufficient or because of lack of geological evidence for the mechanism.

However, Spicer and Herman (2010) used a multivariate paleoclimate proxy known as Climate Leaf Analysis Multivariate Program (CLAMP) to reconstruct temperature, humidity, and precipitation for the Cretaceous Arctic. CLAMP is a statistical tool and the assumption of the program is that no single feature of a leaf can correlate with a specific climatic parameter. Rather, it is the sum of features that define leaf architecture, and that architecture is a function of adaptation to climate. Spicer and Herman's (2010) temperature values compared well with the previous values presented in earlier work. Their estimates for precipitation throughout the ancient Arctic also suggested ample rainfall, though their available samples were limited. Their CLAMP estimate of Arctic annual average relative humidity is high, ranging from 70% to 85%. Spicer and Herman (2010) suggested, based on the paleobotanical record, that the modeled high values for humidity, precipitation, and warm temperatures supported the existence of a persistent polar cloud cap that helped maintain high temperatures through the polar winter. Going further, others (Kump and Pollard, 2008; Upchurch et al., 2015) suggested that altering the cloud cover properties accounts for warming the Arctic while avoiding overheating the tropical regions during the Cretaceous.

With respect to the model of moving heat to the higher latitudes via latent oceanic heat (Herman and Spicer, 1996; Johnson et al., 1996), an interesting historical parallel regarding ocean heat transport warming the Cretaceous Arctic stems from the 19th-century fascination with the hypothesized Open Polar Sea—a fascination which motivated the famed Arctic explorations of the that century. Mercator's 1595 map of the Arctic shows a symmetrical arrangement of land about the pole with four large rivers flowing north and draining into an open northern ocean. Despite the fact that no one into the 19th century had seen this open polar sea, the idea of it was carried through on maps. How this polar sea stayed open was the subject of discussion, even more than whether it even existed.

By the mid-1800s the field of oceanography was emerging, and one person in particular, Silas Bent, became an advocate of circumpolar oceanic circulation that would at least contribute to keeping the Open Polar Sea open (Bent, 1869). The model was based on the hypothesized course of the Gulf Stream in the Atlantic Ocean and the Kuro Siwa in the Pacific Ocean, both warm currents moving northward. Where the Kuro Siwa flowed was unknown at the time but Bent made analogies to circulation in the air and in the bodies of animals. From this analogy, he proposed that these two giant oceanic currents pumped heat north to warm the Arctic (Figure 8.5).

Understanding the relative roles of these various mechanisms (oceanic heat transport, continental paleogeography, increased atmospheric greenhouse gases) in shaping

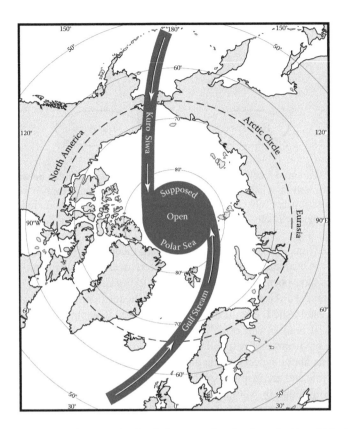

**FIGURE 8.5** Circumpolar oceanic circulation map, showing how the Open Polar Sea might exist. (Redrawn from Bent, S. 1869. *An Address Delivered before the St. Louis Historical Society, December 10, 1868. Upon the Thermometric Gateways to the Pole, Surface Currents of the Ocean, and the Influence of the Latter upon the Climate of the World.* St. Louis: RP Studley & Company.)

climate is critical. It is a truism that Earth system models fail to accurately recreate the past, though these models are used to predict a future greenhouse world. This lack of ability to account for the Cretaceous record suggests there remain fundamental gaps in our understanding of processes and feedbacks that control high-latitude climate. Similarly, new mechanisms have been proposed to account for Cretaceous warm high latitudes. One mechanism recognizes the presence of expansive polar forests that would result in changes to surface albedo (Otto-Bliesner and Upchurch, 1997; DeConto et al., 2000; Zhou et al., 2012). Another calls for enhanced surface methane emission rates and atmospheric concentrations, though the sources for these enhanced emissions vary from wetlands to marine (Ludvigson et al., 2002; Bice et al., 2006; Beerling et al., 2011). Other mechanisms, which center on changes to atmospheric boundary conditions, include the formation of polar stratospheric (Sloan and Pollard, 1998) or convective clouds (Abbot and Tziperman, 2008), changes in cloud microphysical processes related to biological feedbacks (Kump and Pollard, 2008), and a reduction in low cloud amounts through low-level atmospheric destabilization

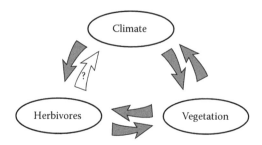

**FIGURE 8.6** Potential feedbacks between biota and climate. The links between vegetation and climate are well established, and the impact of climate on herbivores is also well established. However, it is entirely reasonable that the manipulations of vegetation by herbivores may have influenced climate.

(Poulsen and Zhou, 2013). As intriguing as these mechanisms are, these models have been difficult to evaluate, or when they have been investigated using modern analogs, those analogs have been inappropriate to apply to the Cretaceous.

That the modern Arctic climate plays a significant role in the overall global climate is well established, even if not fully understood (Chapin et al., 2000; McGuire et al., 2006). It is also well established that the structure and function of modern high-latitude environments respond to even subtle changes in climate. Here I discuss a few thoughts about aspects of the ancient Arctic climate, but these are limited to the carbon cycle and are inspired by the presence of the rich ancient later Cretaceous dinosaur record found in Alaska.

Elephants, rhinoceroses, giraffes, and hippopotami—modern megaherbivores—have greater effects on their ecosystem than smaller herbivores (Owen-Smith, 1988), so this robust and still growing record of fossil vertebrates would seem to suggest the presence of some potential feedbacks between the biota and climate (Figure 8.6) during the Cretaceous, particularly given the large body sizes of dinosaurs. Thus, as it is now recognized that data gathered on relationships and feedback loops in modern ecosystems furthers a global understanding of climate, a similar understanding of past regional polar terrestrial processes may help in future global climate modeling.

## METHANE

Modern climatologists investigating Arctic climate variation over the course of the last few hundred years concern themselves with understanding parameters such as how concentrations of greenhouse gases such as $CO_2$ and methane have changed through time, or how soil moisture has varied, or understanding relatively recent past fire dynamics. It is reasonable to assume that the presence of large-bodied vertebrates such as the Cretaceous Arctic dinosaurs similarly provides insight regarding understanding ancient climate.

Because $CO_2$ is one of the most abundant greenhouse gases, and it has a long residence time (low turnover rate) in the atmosphere, significant effort to understand modern climates is spent on understanding $CO_2$ cycles. Similarly, much of the deep time climate discussion has focused on $CO_2$ concentrations.

However, $CO_2$ is just one of the three primary greenhouse gases, nitrous oxide and methane being the other two. As Carl Sagan (1994) so simply pointed out, the presence of methane in the atmosphere is unusual given that the laws of chemistry dictate that methane in the presence of oxygen should be converted to water vapor. The presence of methane in the atmosphere of Earth suggests that it is being introduced at a far greater rate than it can be converted to water vapor. Sagan further suggested that the source must be biological, based on the biochemistry of life and how unstable methane is in an oxygen-rich atmosphere.

A growing body of research on modern climates now clearly demonstrates that methane concentrations are more significant within short-term time intervals (e.g., 20 years; Pacala et al., 2010) and rates of methane production affect patterns of global warming significantly (e.g., Kennett et al., 2003; Forster et al., 2007; Dlugkencky et al., 2011; Isaksen et al., 2011). Presently the highest atmospheric concentrations of methane as well as the highest global amplitudes of seasonal changes in methane concentrations occur in the Arctic, illustrating the importance of the Arctic terrestrial biosphere in understanding the processes for regulating atmospheric methane.

Sources of methane input and their relative importance are still being understood, and range from methane hydrates (Kennett et al., 2003) to thawing of permafrost in the Arctic (Isaksen et al., 2011) and Arctic soils (Jahren et al., 2004) to a variety of animal emissions, particularly domesticated livestock (Johnson and Johnson, 1995; Benchaar et al., 1998; Steinfeld et al., 2006; Franz et al., 2011a,b). Whereas thoughts of the contributions of the latter may produce smirks for some readers, the contribution of livestock, specifically cattle, has justifiably been a serious subject of study relatively recently (e.g., Johnson and Johnson, 1995; Benchaar et al., 1998; Steinfeld et al., 2006). It has been suggested that modern ruminants contribute 17% of the annual emissions of methane to the atmosphere (Dlugokencky et al., 2011).

Herbivores create methane as energy loss during digestion, and the relationship between body mass and percentage of energy loss indicates that as body mass increases, so does methane production (e.g., Franz et al., 2011a,b; Figure 8.7). This pattern holds regardless of whether studying ruminants, small mammals, or reptiles

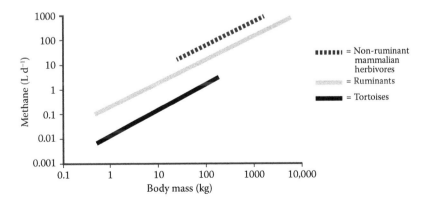

**FIGURE 8.7** Methane output as a function of body size. Notice the similar relationship independent of taxon.

**FIGURE 8.8**   Large footprint attributable to an adult hadrosaur found in the Lower Cantwell Formation, Denali National Park, Alaska. This type of track is the most commonly found track within this Cretaceous rock unit.

(Franz et al., 2011a,b), the former group having the strongest increase in methane production with increasing body size (Franz et al., 2011a). The process can be mitigated in domestic livestock (e.g., Martin et al., 2010), so it is unclear why herbivorous animals have not evolved a means of their own to increase digestive efficiency and reduce methane loss. While this evolutionary question remains unresolved, given that animals today have methane energy losses during digestion, it is reasonable to assume that animals through time experienced similar dietary inefficiencies.

Dinosaurs were abundant during the Cretaceous, and dinosaurs were among the largest-bodied terrestrial animals ever to walk the earth. By virtue of the abundance and considerable body mass of individuals, herbivorous dinosaurs presumably affected their ecosystem in ways that qualified them as keystone species (sensu Paine, 1969).

The largest herbivorous dinosaurs known from Alaska were hadrosaurs and ceratopsians (Chapters 4 and 5). Comparing the relative frequency of hadrosaur-dominated bonebeds to ceratopsian-dominated bonebeds in the Prince Creek Formation of northern Alaska shows that the former outnumber the latter four to one (Chapter 7). Similarly, the footprint occurrences (Figure 8.8) of each group found in correlative rocks within Denali National Park and Preserve in the central Alaska Range also show a dominance by hadrosaurs. Other footprint occurrences exist from Aniakchak National Monument and Preserve, Wrangell–St. Elias National Park and Preserve, and Yukon–Charley Rivers National Preserve, but these records are sparse.

In the matter of large herbivore methane production, the calculated mass of *Edmontosaurus* (Anderson et al., 1985) is assumed to be comparable to that of a modern elephant (Spinage, 1994), approximately 4500 kg. The basic metabolic needs of a hadrosaur and an elephant were, for the sake of argument, taken to be the same.

Similarly, also it is assumed that modern ecosystem productivity values approximate those in the high latitudes of the Cretaceous.

In Uganda, the density of elephants ranges from 2.8 to 3.5 elephants per square kilometer (Spinage, 1994). The state of Alaska is approximately 1,500,000 square kilometers and was roughly assembled by plate movements by the Late Cretaceous (Chapter 3). Hadrosaurs occur across the state (Fiorillo and Parrish, 2004; Fiorillo et al., 2010a,b, 2012, 2014a,b; Gangloff and Fiorillo, 2010). Taking the number three as an approximation for elephants in Uganda and applying it to hadrosaurs in an area the size of Alaska yields a figure of some 500,000 hadrosaurs living in Alaska during any given time within the Late Cretaceous. Not all of Alaska during the Cretaceous was prime hadrosaur habitat, so this estimate of standing population is a maximum. In addition, population data from a very large hadrosaur tracksite in Denali National Park shows that although hadrosaur herds were dominated by the largest size class, approximately 12%–13% of a herd consisted of smaller, immature individuals (Fiorillo et al., 2014a). Thus not all of these animals were fully adult, thereby also reducing the calculation of methane production.

A modern domestic cow (or bull) can be considered to be approximately 450 kg in mass; therefore it would take ten cows to equal the mass of one adult hadrosaur. A cow of this size produces approximately 1500 kg of manure per year (Colorado Comprehensive Nutrient Management Plan Workbook; www.extsoilcrop.colostate.edu/Soils/cnmp/toc.html). One kilogram of cow manure produces approximately 14 liters of methane at 28°C (http://www.habmigern2003.info/biogas/methane-digester.html). If a single cow can produce 21,000 liters of methane per year from manure, then does that mean that hadrosaur manure may have generated 210,000 liters of methane per year? And if there were 500,000 hadrosaurs in Alaska, does that mean these ancient arctic animals annually may have contributed over 105,000,000,000 liters of methane per year from their manure generation to the atmosphere?

Cattle alone are estimated to annually contribute approximately 2% of the methane introduced into the atmosphere (Johnson and Johnson, 1995), though estimates for ruminants as a group can approach 17% (Dlugokencky et al., 2011), which is considered a potentially significant contribution to global warming (Johnson and Johnson, 1995; Benchaar et al., 1998). It may be that hadrosaurs played a similar role to ruminants in producing methane emissions that impacted global climate during the Cretaceous (Figure 8.9). That role for hadrosaurs was likely even more significant if one considers the global population of the taxon rather than restricting the discussion to Arctic forms. While the exact numbers presented here may be suspect, given that this discussion assumes that hadrosaurs had similar digestive processes to ruminants, the scale of the proposed quantities of emitted methane from Arctic hadrosaurs is nonetheless illuminating.

## FIRE

A second parameter that concerns modern climatologists is fire system dynamics. There are two basic types of wildfire—crown fires and surface fires. The crown of a tree refers to the branches and leaves of an individual tree, while a canopy is the interconnected cover of individual crowns. Crowns provide the fuel for crown fires. Surface fires burn fuels accumulated on the ground. Crown fires are more intense, and the heat

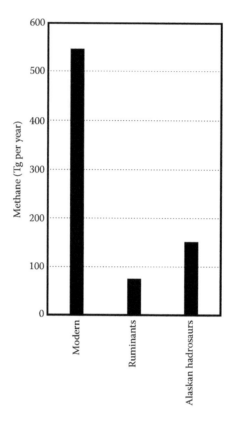

**FIGURE 8.9** Estimate of methane emission by Alaskan hadrosaurs compared to modern methane emissions, and the contribution by ruminants to annual methane emissions. (Modern and ruminant emission values modified from Wilkinson, D.M. et al. 2012. Could methane produced by sauropod dinosaurs have helped drive Mesozoic climate warmth? *Current Biology* 22:R292–R293.)

generated is greater than surface fires. Primary factors that can drive the transition from surface fire to crown fire include wind and heat released from surface fuels (Scott and Reinhardt, 2001). The behavior of individual fires is determined by intensity, rate of spread, and heat release per unit area (Scott and Reinhardt, 2001).

Wildfire is a contributor to the landscape processes that form a dynamic boundary contributing to climate. The disturbance and destabilization of the landscape caused by fire impacts the partitioning of incoming solar radiation and directly effects the introduction of carbon into the atmosphere (Crutzen and Andreae, 1990; Crutzen and Goldammer, 1993; Clark, 1997). Fire is the process that links the biosphere, hydrosphere, and atmosphere, and is a naturally occurring phenomenon that has been in place since the earliest land plants (Scott and Glasspool, 2006). In addition, recognizing that forest fires can significantly increase the availability of sediment and runoff from a fire-generated destabilized landscape (e.g., Meyer and Wells, 1997). Cretaceous fires may even have contributed to the formation of some important dinosaur-bearing deposits (Brown et al., 2013).

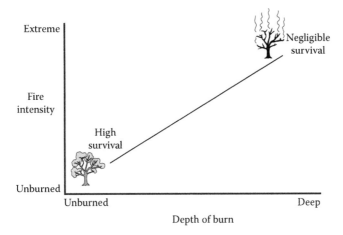

**FIGURE 8.10**  Illustration of variation in fire severity and the impact on the survivorship of biota.

The frequency of fires results from a combination of factors such as vegetation type, amount of fuel buildup, topography, and weather. The intensity and severity of individual fires determines survivorship of flora (Figure 8.10), while the frequency is related to atmospheric humidity as well as the moisture content of vegetation (Uhl et al., 1988). Interestingly, many populations of vertebrates show little if any negative effects from fire-affected landscapes (Lyon et al., 2000).

Burns can restructure ecosystems at least for the short term with respect to geologic time (Figure 8.11). There is a feedback loop between fire and climate, as fires contribute gases and particulates to the atmosphere (Figure 8.12), while climate plays a role in fire frequency. Vegetation type can also play a role in fire frequency and intensity. For example, there are differences in albedo in the modern boreal forest biome between boreal conifer stands, boreal deciduous stands, and herbaceous open ground (McGuire and Chapin, 2006). During the summer within the modern boreal forest biome, conifer stands have half the albedo of the other two types of vegetated areas (McGuire and Chapin, 2006). The absorbed heat within conifer stands creates thermal convection, which in turn impacts thunderstorm activity and resulting lightning frequency (McGuire and Chapin, 2006). As lightning is the non-human cause for igniting wildfires (Brown, 2000; Kasischke et al., 2006), significant changes in vegetation would impact lightning strike frequency and resulting wildfire frequency. These changes in the release of carbon to the atmosphere on a global scale would then have some role in altering climate, and therefore this linkage between changes in fire frequency and climate is of primary interest to modern climatologists. In addition, within the northern forest biome where cold and dry dominate and therefore limit carbon cycling, fire is the main driver for vegetation recycling compared to more moist and warm environments where decay commonly occurs (Brown, 2000).

Spicer (2003) compared the relative abundance of charcoal in the Prince Creek Formation with that of the older Cretaceous Nanushuk Formation. He noted an increase in the frequency and thickness of charcoal horizons, which he suggested was

**FIGURE 8.11**   Photograph taken in 2010, just outside of Circle, Alaska, of the aftermath of the Crazy Mountain Fire Complex. The fire started in August, 2009 and burned off much of the mature forest, opening the way for recolonization by herbaceous flora.

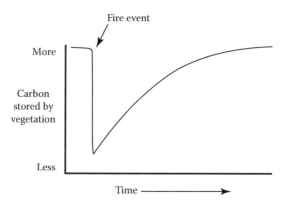

**FIGURE 8.12**   Schematic illustration of carbon release to atmosphere after a fire event.

a result of a change in wildfire patterns between the two rock units. Fires recorded within the Nanushuk Formation had lower temperatures and shorter duration than the fires that burned during the deposition of the Prince Creek Formation. Spicer further speculates that the Prince Creek organic-rich beds represent swamps rather than bogs, and therefore more woody material may have been available to feed wildfires. Implicit in this discussion is a change in fire intensity, which is a metric describing the energy released during a wildfire, as well as fire severity, which is a separate metric describing the damage to an ecosystem caused by fire (Keeley, 2009).

Prince Creek Formation pollen data indicate a mix of woodlands and open cover (Fiorillo et al., 2010a; Flaig et al., 2013), which, pertinent to this discussion, lends itself to the question of how these open areas were maintained. The relative roles of grazing, fire, and climate affect the relationship between woody vegetation and herbaceous vegetation (Fuhlendorf et al., 2008).

Elephants are often viewed as analogs for large-bodied herbivorous dinosaurs. With respect to modern ecosystems containing elephants, such as the Serengeti, the opening of woodlands seems to have been driven by fire, not elephants (Dublin et al., 1990). Fires opened the woodlands, while elephants helped maintain the grasslands. Dublin et al. (1990) showed that fire could not suppress seedling regeneration to adequate levels to keep grasslands open, but elephants could suppress seedling regeneration through foraging.

The abundance of hadrosaurs throughout the Late Cretaceous of Alaska was not trivial to the dynamics of the ecosystem and the climate system (Figure 8.5). The jaw mechanics of hadrosaurs and ceratopsians indicate highly efficient processing of plant food (Weishampel, 1984; Weishampel and Norman, 1989). Hadrosaur coprolites and tooth microwear suggest that the hadrosaur diet consisted primarily of conifers (Chin and Gill, 1996; Fiorillo, 2011). Given the large body size of adult herbivorous dinosaurs and their efficiency in plant food consumption, the interaction between fire intensity and severity, vegetation, and dinosaurs in the Cretaceous of Alaska in maintaining open spaces was likely similar to that of modern elephants.

In their study in the southern plains of the relationship between fire, grazing, and tree–grass interactions, Fuhlendorf et al. (2008) noted a feedback loop between grazing and open areas such that increased herbaceous biomass increased fire intensity, curtailing the spread of the conifer *Juniperus*. Similarly, in the Cretaceous of Alaska, once open areas were established, as recorded by the fossil pollen record (Fiorillo et al., 2010a; Flaig et al., 2013), the increase in herbaceous biomass would have resulted in the fire intensity inhibiting the spread of conifers.

To speculate further about this feedback relationship, because Arctic populations fluctuate (Chapter 7), it is reasonable to think that changes in the Arctic dinosaur populations occurred over time. These changes may have led to changes not only of boundaries between vegetative patches within the Arctic Cretaceous terrestrial ecosystem but also the relative percentages of vegetative patches on the landscape. Changes in vegetative cover led to changes in albedo and influenced the storm cycles and lightning frequencies, as well as possibly altering the frequency of crown verses surface fires. Lightning frequency would change patterns of influx of carbon into the atmosphere, and may have contributed to changes in global climate.

After the Cretaceous-Paleogene extinction, no large mammalian herbivores supplanted the dinosaurs until millions of years later. A slight increase in mean annual temperature at sea level occurred during the Paleocene of Alaska (Spicer and Parrish, 1990), which implies a commensurate change within the terrestrial ecosystem. The importance of forests as regulators on warm climatic intervals in the Cretaceous was suggested (Otto-Bliesner and Upchurch, 1997), so the loss of animals such as the hadrosaurs in the ecosystem, which were consumers of conifers rather than herbaceous plants (Chin and Gill, 1996; Fiorillo, 2011), allowed the forests to expand at the expense of the herbaceous open areas. Admittedly speculative, this scenario would change albedo within this northern terrestrial Arctic ecosystem, as well as change the fire regime and the feedback between the biosphere and climate.

Studies of modern wildfires have shown fires are more frequent in the tropics and subtropics, and the timing of fire season occurs in the transition from dry tropical winter to tropical summer (Harrison et al., 2010). Sedimentological analysis of the Prince Creek Formation shows that these rocks were deposited under seasonal wet–dry conditions (Fiorillo et al., 2010a,b; Flaig et al., 2011, 2013, 2014). Given that this landscape, by virtue of the paleolatitude, experienced dramatic annual changes in light regime, it is reasonable to speculate that the drier part of the year was in winter when plants are dormant due to reduced sunlight. The more pronounced portion of the fire cycle corresponded to the spring–summer transition.

The abundance of herbivorous dinosaurs, specifically hadrosaurs, in the Cretaceous of Alaska, and their efficiency in mastication must have played a role in shaping ecosystem structure. Given that role in shaping the structure of the ecosystem, dinosaurs also contributed to climate feedback, and therein lies the link between local habitat modification and a more global consequence during the Cretaceous.

## SUMMARY

Dinosaurs are Earth's largest terrestrial animals. Their widespread occurrence and abundance in Alaska is evidence that the Cretaceous Arctic world manifested significant biological productivity. Alaskan dinosaurian megaherbivores were gregarious, so in addition to their individual needs and ecological impacts, the social behavior of these animals also likely influenced their ecosystem.

The relationship between vertebrates and climate is not fully understood, but it is recognized that agricultural animals contribute a significant measurable quantity of methane, an important greenhouse gas, to the atmosphere. Given their large body size and abundance, megaherbivorous dinosaurs likely contributed significant quantities of methane to the atmosphere during the Cretaceous.

As dinosaurs inhabited the globe, the sum of their impact *in toto* could have influenced Cretaceous climate. The presence of dinosaurian megaherbivores and sediments in which their fossils are found demonstrates water was at least seasonally abundant, most readily available in warmer months. The wet–dry seasonality of Alaska during the Cretaceous may also have experienced a fire history similar to that observed in the modern world, where fires were more frequent in the transition from a dry winter to a wet summer. In combination with these fires, the dinosaurian megaherbivores likely helped maintain open areas in the vegetative landscape.

## REFERENCES

Abbot, D.S. and E. Tziperman. 2008. A high latitude convective cloud feedback and equable climates. *Quarterly Journal of the Royal Meteorological Society* 134:165–185.

Anderson, J.F., A. Hall-Martin, and D.A. Russell. 1985. Long-bone circumference and weight in mammals, birds and dinosaurs. *Journal of Zoology* 207:53–61.

Bailey, I.W. and E.W. Sinnott. 1915. A botanical index of Cretaceous and Tertiary climates. *Science* 41:831–834.

Bailey, I.W. and E.W. Sinnott. 1916. The climatic distribution of certain types of angiosperm leaves. *American Journal of Botany* 3:24–29.

Barrell, J. 1916. Influence of Silurian-Devonian climates on the rise of air-breathing vertebrates. *Bulletin of the Geological Society of America* 27:387–436.

Barron, E.J., W.W. Hay, and S. Thompson. 1989. The hydrologic cycle: A major variable during Earth history. *Palaeogeography, Palaeoclimatology, Palaeoecology* 75:157–174.

Beechey, F.W. 1831. *Narrative of a Voyage to the Pacific, to Co-operate with the Polar Expeditions: Performed in His Majesty's Ship Blossom.* London: Henry Colburn and Richard Bentley, 2 volumes.

Beerling, D.J.D., A.A. Fox, D.S.D. Stevenson, and P.J.P. Valdes. 2011. Enhanced chemistry-climate feedbacks in past greenhouse worlds. *Proceedings of the National Academy of Sciences* 108:9770–9775.

Benchaar, C., J. Rivest, C. Pomar, and J. Chiquette. 1998. Prediction of methane production from dairy cows using existing mechanistic models and regression equations. *Journal of Animal Science* 76:617–627.

Bent, S. 1869. *An Address Delivered before the St. Louis Historical Society, December 10, 1868. Upon the Thermometric Gateways to the Pole, Surface Currents of the Ocean, and the Influence of the Latter upon the Climate of the World.* St. Louis: RP Studley & Company.

Bice, K.L., D. Birgel, P.A. Meyers, K.A. Dahl, K.-U. Hinrichs, and R.D. Norris. 2006. A multiple proxy and model study of Cretaceous upper ocean temperatures and atmospheric $CO_2$ concentrations. *Paleoceanography* 21(2), PA2002, doi:10.1029/2005PA001203.

Brown, J.K. 2000. Introduction and fire regimes. In *Wildland Fire in Ecosystems: Effects of Fire on Flora*, eds. J.K. Brown and J.K. Smith, 1–7. Ogden, UT: U.S. Forest Service, Rocky Mountain Research Station. United States Department of Agriculture Forest Service General Technical Report RMRS-GTR-42 2.

Brown, S.A.E., M.E. Collinson, and A.C. Scott. 2013. Did fire play a role in formation of dinosaur-rich deposits? An example from the Late Cretaceous of Canada. *Palaeobiodiversity and Palaeoenvironments* 93:317–326.

Buckland, W.M. 1831. On the occurrence of the remains of Elephants, and Quadrupeds, in the cliffs of frozen mud, in Eschscholtz Bay, within Beering's Strait, and in other distant parts of the shores of the Arctic Seas. In *Narrative of a Voyage to the Pacific, to Co-Operate with the Polar Expeditions: Performed in His Majesty's Ship Blossom*, ed. F.W. Beechey. London: Henry Colburn and Richard Bentley. Volume 2 appendix.

Chaloner, W.G. and G.T. Crebin. 1990. Do fossil plants give a climatic signal? *Journal of the Geological Society* 147:343–350.

Chapin, F.S., A.D. McGuire, J. Randerson, R. Pielke, D. Baldocchi, S.E. Hobbie, N. Roulet, W. Eugster, E. Kasischke, E.B. Rastetter, and S.A. Zimov. 2000. Arctic and boreal ecosystems of western North America as components of the climate system. *Global Change Biology* 6(S1):211–223.

Chin, K. and B.D. Gill. 1996. Dinosaurs, dung beetles, and conifers: Participants in a Cretaceous food web. *Palaios* 11:280–285.

Clarke, A. and K.J. Gaston. 2006. Climate, energy and diversity. *Proceedings of the Royal Society B* 273:2257–2266.

Clark, J.S. 1997. An introduction to sediment records of biomass burning. In *Sediment Records of Biomass Burning and Global Change*, eds. J.S. Clark, H. Cachier, J.G. Goldammer, and B. Stocks, 1–5. Berlin: NATO ASI Series I51, Springer-Verlag.

Crutzen, P.J. and M.O. Andreae. 1990. Biomass burning in the tropics: Impact on atmospheric chemistry and biogeochemical cycles. *Science* 250:1669–1678.

Crutzen, P.J. and J.G. Goldammer (eds.). 1993. *Fire in the Environment: The Ecological, Atmospheric and Climatic Importance of Vegetation Fires.* New York: John Wiley & Sons.

Cuvier, G. 1825. *Recherches sur les ossements fossils, ou l'on retablit les characteres des plusieurs animaux dont les revolutions du globe ont detruit les especes.* 3rd edition. Paris: Dufour & d'Ocagne.

DeConto, R.M., E.C. Brady, J. Bergengren, S.L. Thompson, D. Pollard, and W.W. Hay. 2000. Late Cretaceous climate, vegetation, and ocean interactions. In *Warm Climates in Earth History*, eds. B.T. Huber, K.G. Macleod, and S.L. Wing, 275–297. Cambridge, UK: Cambridge University Press.

DeConto, R.M., W.W. Hay, S.L., Thompson, and J. Bergengren, 1999. Late Cretaceous climate and vegetation interactions: Cold continental interior paradox. In *Evolution of the Cretaceous Ocean-Climate System*, eds. E. Barrera and C.C. Johnson, 391–406. Boulder, CO: Geological Society of America Special Paper 332.

Dlugokencky, E., E.G. Nisbet, R. Fisher, and D. Lowry. 2011. Global atmospheric methane: Budget, changes and dangers. *Philosophical Transactions of the Royal Society A* 369:2058–2072.

Donnadieu, Y., R. Pierrehumbert, R. Jacob, and F. Fluteau. 2006. Modelling the primary control of paleogeography on Cretaceous climate. *Earth and Planetary Science Letters* 248:426–437.

Dublin, H.T., A.R.E. Sinclair, and J. McGlade. 1990. Elephants and fire as causes of multiple stable states in the Serengeti-Mara woodlands. *Journal of Animal Ecology* 59:147–1164.

Evans, K.L., J.J.D. Greenwood, and K.J. Gaston. 2005. Dissecting the species-energy relationship. *Proceedings of the Royal Society B* 272:2155–2163.

Fiorillo, A.R. 2011. Microwear patterns on the teeth of northern high latitude hadrosaurs with comments on microwear patterns in hadrosaurs as a function of latitude and seasonal ecological constraints. *Palaeontologia Electronica* 14:20A, 17p.

Fiorillo, A.R., T.L. Adams, and Y. Kobayashi. 2012. New sedimentological, palaeobotanical, and dinosaur ichnological data on the palaeoecology of an unnamed Late Cretaceous rock unit in Wrangell–St. Elias National Park and Preserve, Alaska, USA. *Cretaceous Research* 37:291–299.

Fiorillo, A.R., F. Fanti, C. Hults, and S.T. Hasiotis. 2014b. New ichnological, paleobotanical and detrital zircon data from an unnamed rock unit in Yukon–Charley Rivers National Preserve (Cretaceous: Alaska): Stratigraphic implications for the region. *Palaios* 29:16–26.

Fiorillo, A.R., S.T. Hasiotis, and Y. Kobayashi. 2014a. Herd structure in Late Cretaceous polar dinosaurs: A remarkable new dinosaur tracksite, Denali National Park, Alaska, USA. *Geology* 42:719–722.

Fiorillo, A.R., S.T. Hasiotis, Y. Kobayashi, and C.S. Tomsich. 2009. A pterosaur manus track from Denali National Park, Alaska Range, Alaska, USA. *Palaios* 24:466–472.

Fiorillo, A.R., P.J. McCarthy, and P.P. Flaig. 2010b. Taphonomic and sedimentologic interpretations of the dinosaur-bearing Upper Cretaceous Strata of the Prince Creek Formation, Northern Alaska: Insights from an ancient high-latitude terrestrial ecosystem. *Palaeogeography, Palaeoclimatology, Palaeoecology* 295:376–388.

Fiorillo, A.R., P.J. McCarthy, P.P. Flaig, E. Brandlen, D.W. Norton, P. Zippi, L. Jacobs, and R.A. Gangloff. 2010a. Paleontology and paleoenvironmental interpretation of the Kikak-Tegoseak Quarry (Prince Creek Formation: Late Cretaceous), northern Alaska: A multi-disciplinary study of a high-latitude ceratopsian dinosaur bonebed. In *New Perspectives on Horned Dinosaurs*, eds. M.J. Ryan, B.J. Chinnery-Allgeier, and D.A. Eberth, 456–477. Bloomington: Indiana University Press.

Fiorillo, A.R. and J.T. Parrish. 2004. The first record of a Cretaceous dinosaur from western Alaska. *Cretaceous Research* 25:453–458.

Flaig, P.P., A.R. Fiorillo, and P.J. McCarthy. 2014. Dinosaur-bearing hyperconcentrated flows of Cretaceous Arctic Alaska: Recurring catastrophic event beds on a distal paleopolar coastal plain. *Palaios* 29:594–611.

Flaig, P.P., P.J. McCarthy, and A.R. Fiorillo. 2011. A tidally influenced, high-latitude coastal-plain: The upper Cretaceous (Maastrichtian) Prince Creek Formation, North Slope, Alaska. In *From River to Rock Record: The Preservation of Fluvial Sediments and Their Subsequent Interpretation*, eds. S.K. Davidson, C.P. North, and S. Leleu, 233–264. Tulsa: SEPM Special Publication 97.

Flaig, P.P., P.J. McCarthy, and A.R. Fiorillo. 2013. A tidally influenced, high-latitude alluvial/coastal plain: The Late Cretaceous (Maastrichtian) Prince Creek Formation, North Slope, Alaska. In *New Frontiers in Paleopedology and Terrestrial Paleoclimatology—Paleosols and Soil Surface Analog Systems*, eds. S.G. Driese, L.C. Nordt, and P.J. McCarthy, 179–230. Tulsa: SEPM Special Publication 10.

Forster, P., V. Ramaswamy, P. Artaxo, T. Berntsen, R. Betts, D.W. Fahey, J. Haywood et al. 2007. In *Changes in Atmospheric Constituents and in Radiative Forcing, in Climate Change 2007: The Physical Science Basis. Contribution of Working Group I to the Fourth Assessment Report of the Intergovernmental Panel on Climate Change*, eds. S. Solomon, D. Qin, M. Manning, Z. Chen, M. Marquis, K.B. Averyt, M. Tignor, and H.L. Miller, 129–134. Cambridge, UK: Cambridge University Press.

Franklin, B. 1767. Benjamin Franklin to George Croghan, August 5, 1767. Letter. In *Memoirs of the Life and Writings of Benjamin Franklin, LL.D. F.R.S. & c*, quarto edition, ed. W.T. Franklin, London: Henry Colburn, 1818, III, 366–367.

Franz, R., C.R. Soliva, M. Kreuzer, J.-M. Hatt, S. Furrer, J. Hummel, and M. Clauss. 2011a. Methane output of tortoises: Its contribution to energy loss related to herbivore body mass. *PLoS ONE* 6(3):e17628. doi:10.1371/journal.pone.0017628.

Franz, R., C.R. Solivia, M. Kreuzer, J. Hummel, and M. Clauss. 2011b. Methane output of rabbits (*Oryctolagus cuniculus*) and guinea pigs (*Cavia porcellus*) fed a hay-only diet: Implications for the scaling of methane production with body mass in non-ruminant mammalian herbivores. *Comparative Biochemistry and Physiology Part A* 158:177–181.

Fuhlendorf, S.D., S.A. Archer, F.E. Smeins, D.M. Engle, and C.A. Taylor, Jr. 2008. The combined influence of grazing, fire, and herbaceous productivity on tree-grass interaction. In *Western North American Juniperus Communities: A Dynamic Vegetation Type*, ed. O.W. Van Auken, 219–238. New York, NY: Springer.

Gangloff, R.A. and A.R. Fiorillo. 2010. Taphonomy and paleoecology of a bonebed from the Prince Creek Formation, North Slope, Alaska. *Palaios* 25:299–317.

Guthrie, R.D. 1990. *Frozen Fauna of the Mammoth Steppe: The Story of Blue Babe*. Chicago, IL: University of Chicago Press.

Harrison, S.P., J.R. Marlon, P.J. Bartlein. 2010. Fire in the earth system. In *Changing Climates, Earth Systems and Society*, ed. J. Dodson, 21–48. Dordrecht: Springer.

Hedeen, S. 2008. *Big Bone Lick: The Cradle of American Paleontology*. Lexington, KY: University of Kentucky Press.

Herman, A.B. and R.A. Spicer. 1996. Paleobotanical evidence for a warm Cretaceous Arctic Ocean. *Nature* 380:330–333.

Herman, A.B., R.A. Spicer, and T.E.V. Spicer. 2016. Environmental constraints on terrestrial vertebrate behavior and reproduction in the high Arctic of the Late Cretaceous. *Palaeogeography, Palaeoclimatology, Palaeoecology.* 441:317–338.

Isaksen, I.S.A., M. Gauss, G. Myhre, K.M. Walter Anthony, and C. Ruppel. 2011. Strong atmospheric chemistry feedback to climate warming from Arctic methane emissions. *Global Biogeochemical Cycles* 25:GB2002, doi:10.1029/2010GB003845.

Jahren, A.H., B.A. LePage, and S.P. Werts. 2004. Methanogenesis in Eocene Arctic soils inferred from $\delta^{13}C$ of tree fossil carbonates. *Palaeogeography, Palaeoclimatology, Palaeoecology* 214:347–358.

Jenkyns, H.C., A. Forster, S. Schouten, and J.S.S. Damste. 2004. High temperatures in the Late Cretaceous Arctic Ocean. *Nature* 432:888–892.

Johnson, C.C., E.J. Barron, E.G. Kauffman, M.A. Arthur, P.J. Fawcett, and M.K. Yasuda. 1996. Middle Cretaceous reef collapse linked to ocean heat transport. *Geology* 24:376–380.

Johnson, K.A. and D.E. Johnson. 1995. Methane emissions from cattle. *Journal of Animal Science* 73:2483–2492.

Kasischke, E.S., T.S. Rupp, and D.L. Verbyla. 2006. Fire trends in the Alaskan boreal forest. In *Alaska's Changing Boreal Forest*, eds. F.S. Chapin, III, M.W. Oswood, K. Van Cleve, L.A. Viereck, and D.L. Verbyla, 285–301. New York, NY: Oxford University Press.

Keeley, J.E. 2009. Fire intensity, fire severity and burn severity: A brief review and suggested usage. *International Journal of Wildland Fire* 18:116–126.

Kennett, J.P., K.G. Cannariato, I.L. Hendy, and R.J. Behl. 2003. *Methane Hydrates in Quaternary Climate Change: The Clathrate Gun Hypothesis*. Washington, DC: American Geophysical Union.

Kump, L.R. and D. Pollard. 2008. Amplification of Cretaceous warmth by biological cloud feedbacks. *Science* 320:195–195.

Ludvigson, G., L.A. González, R.A. Metzger, B.J. Witzke, R.L. Brenner, A.P. Murillo, and T.S. White. 1998. Meteoric sphaerosiderite lines and their use for paleohydrology and paleoclimatology. *Geology* 26:1039–1042.

Ludvigson, G.A., L.A. González, D.F. Ufnar, B.J. Witzke, and B.L. Brenner. 2002. Methane fluxes from mid-Cretaceous wetland soils: Insights gained from carbon and oxygen isotopic studies of sphaerosiderites in paleosols. *Geological Society of America Abstracts with Programs* 34(6):212.

Lyon, L.J., M.H. Huff, E.S. Telfer, D.S. Schreiner, and J.K. Smith. 2000. Fire effects on animal populations. In *Wildland Fire in Ecosystems: Effects of Fire on Fauna*, ed. J.K. Smith, 25–34. Ogden UT: USDA Forest Service. United States Department of Agriculture Forest Service, General Technical Report RMRS-GTR-42 1.

Manabe, S. and R.T. Wetherald. 1980. On the distribution of climate change resulting from an increase in $CO_2$ content of the atmosphere. *Journal of the Atmospheric Sciences* 37:99–118.

Martin, C., D.P. Morgavi, and M. Doreau. 2010. Methane mitigation in ruminants: From microbe to the farm scale. *Animal* 4:351–365.

Matthew, W.D. 1915. Climate and evolution. *Annals of the New York Academy of Sciences* 24:171–318.

McGuire, A.D. and F.S. Chapin, III. 2006. Climate feedbacks in the Alaskan Boreal Forest. In *Alaska's Changing Boreal Forest*, eds. F.S. Chapin, III, M.W. Oswood, K. Van Cleve, L.A. Viereck, and D.L. Verbyla, 309–322. New York, NY: Oxford University Press.

McGuire, A.D., F.S. Chapin, III, J.E. Walsh, and C. Wirth. 2006. Integrated regional changes in arctic climate feedbacks: Implications for the global climate system. *Annual Review of Environment and Resources* 31:61–91.

Meyer, G.A. and S.G. Wells. 1997. Fire-related sedimentation events on alluvial fans, Yellowstone National Park, USA. *Journal of Sedimentary Research* 67:776–791.

Otto-Bliesner, B. and G.R. Upchurch. 1997. Vegetation-induced warming of high-latitude regions during the late Cretaceous period. *Nature* 385:804–807.

Owen-Smith, R.N. 1988. *Megaherbivores: The Influence of Very Large Body Size on Ecology*. Cambridge, UK: Cambridge University Press.

Pacala, S.W., C. Breidenich, P.G. Brewer, I. Fung, M.R. Gunson, G. Heddle, B. Law, G. Marland, K. Paustian, M. Prather, J.T. Randerson, P. Tans, and S.C. Wofsy. 2010. *Verifying Greenhouse Gas Emissions*. Washington, DC: The National Academies Press.

Paine, R.T. 1969. A note on trophic complexity and community stability. *The American Naturalist* 103:91–93.

Poulsen, C.J. and J. Zhou. 2013. Sensitivity of Arctic climate variability to mean state: Insights from the Cretaceous. *Journal of Climate* 26:7003–7022.

Sagan, C. 1994. *Pale Blue Dot: A Vision of the Human Future in Space.* New York, NY: Random House.

Salazar Jaramillo, S., P.J. McCarthy, T.P. Trainor, S.J. Fowell, and A.R. Fiorillo. 2015. Origin of clay minerals in alluvial paleosols, Prince Creek Formation, North Slope, Alaska, USA: Influence of volcanic ash on pedogenesis in the Late Cretaceous Arctic. *Journal of Sedimentary Research* 85:192–208.

Salazar-Jaramillo, S., S.J. Fowell, P.J. McCarthy, J.A. Benowitz, M.G. Śliwiński, and C.S. Tomsich. 2016. Terrestrial isotopic evidence for a Middle-Maastrichtian warming event from the lower Cantwell Formation, Alaska. *Palaeogeography, Palaeoclimatology, Palaeoecology* 441:360–376.

Scott, A.C. and I.J. Glasspool. 2006. The diversification of Paleozoic fire systems and fluctuations in atmospheric oxygen concentrations. *Proceedings of the National Academy of Sciences* 103:10861–10865.

Scott, J.H. and E.D. Reinhardt. 2001. Assessing crown fire potential by linking models of surface and crown fire behavior. United States Department of Agriculture Forest Service Research Paper RMRS-RP-29:1–59 p.

Sloan, L.C. and D. Pollard. 1998. Polar stratospheric clouds: A high latitude warming mechanism in an ancient greenhouse world. *Geophysical Research Letters* 25:3517–3520.

Spicer, R.A. 2003. Changing climate and biota. In *The Cretaceous World*, eds. P. Skelton, R.A. Spicer, S. Kelley, and I. Gilmour, 85–162. Cambridge, UK: Cambridge University Press.

Spicer, R.A. and A.B. Herman. 2010. The Late Cretaceous environment of the Arctic: A quantitative reassessment based on plant fossils. *Palaeogeography, Palaeoclimatology, Palaeoecology* 295:423–442.

Spicer, R.A. and J.T. Parrish. 1986. Paleobotanical evidence for cool North Polar climates in middle Cretaceous (Albian-Cenomanian) time. *Geology* 14:703–706.

Spicer, R.A. and J.T. Parrish. 1990. Late Cretaceous—Early Tertiary palaeoclimates of northern high latitudes: A quantitative view. *Journal of the Geological Society* 147:329–341.

Spinage, C. 1994. *Elephants.* London, UK: T & AD Poyser Ltd.

Steinfeld, H., P. Gerber, T. Wassenaar, V. Castel, M. Rosales, and C. de Haan. 2006. *Livestock's Long Shadow: Environmental Issues and Options.* Rome, IT: Food and Agriculture Organization of the United Nations.

Suarez, C.A., P.P. Flaig, G.A. Ludvigson, L.A. González, R. Tian, H. Zhou, P.J. McCarthy, D.A. Van der Kolk, and A.R. Fiorillo. 2016. Reconstructing the paleohydrology of a Cretaceous Alaskan paleopolar coastal plain from stable isotopes of bivalves. *Palaeogeography, Palaeoclimatology, Palaeoecology* 441:339–351.

Suarez, C.A., G.A. Ludvigson, L.A. González, A.R. Fiorillo, P.P. Flaig, and P.J. McCarthy. 2013. Use of multiple oxygen isotope proxies for elucidating Arctic Cretaceous Palaeo-hydrology. In *Isotopic Studies in Cretaceous Research*, eds. A.-V. Bojar, M.C. Melinte-Dobrinescu, and J. Smit, 185–202. London, UK: Geological Society, Special Publications 382.

Suarez, M.B., L.A. González, and G.A. Ludvigson. 2011. Quantification of a greenhouse hydrologic cycle from equatorial to polar latitudes: The mid-Cretaceous water bearer revisited. *Palaeogeography, Palaeoclimatology, Palaeoecology* 307:301–312.

Thomson, K. 2008. *The Legacy of the Mastodon: The Golden Age of Fossils in America.* New Haven, CT: Yale University Press.

Tomsich, C.S., P.J. McCarthy, S.J. Fowell, and D. Sunderlin. 2010. Paleofloristic and paleoenvironmental information from a Late Cretaceous (Maastrichtian) flora of the lower Cantwell Formation near Sable Mountain, Denali National Park, Alaska. *Palaeogeography, Palaeoclimatology, Palaeoecology* 295:389–408.

Ufnar, D.F., L.A. González, G.A. Ludvigson, R.L. Brenner, and B.J. Witzke. 2001. Stratigraphic implications of meteoric sphaerosiderite δ¹⁸O values in paleosols in the Cretaceous (Albian) Boulder Creek Formation, NE British Colombia Foothills, Canada. *Journal of Sedimentary Research* 71:1017–1028.

Ufnar, D.F., L.A. González, G.A. Ludvigson, R.L. Brenner, and B.J. Witzke. 2002. The mid-Cretaceous water bearer: Isotope mass balance quantification of the Albian hydrologic cycle. *Palaeogeography, Palaeoclimatology, Palaeoecology* 188:51–71.

Ufnar, D.F., G.A. Ludvigson, L.A. González, R.L. Brenner, and B.J. Witzke. 2004. High latitude meteoric δ¹⁸O compositions: Paleosol siderite in the middle Cretaceous Nanushuk Formation, North Slope, Alaska. *Geological Society of America Bulletin* 116:463–473.

Uhl, C.K., J.B. Kaufman, and D.L. Cummings. 1988. Fire in the Venezuelan Amazon 2: Environmental conditions necessary for forest fires in the evergreen rainforest of Venezuela. *Oikis* 53:176–184.

Upchurch, G.R., J. Kiehl, C. Shields, J. Scherer, and C. Scotese. 2015. Latitudinal temperature gradients and high-latitude temperatures during the latest Cretaceous: Congruence of geologic data and climate models. *Geology* 43:683–686.

Weishampel, D.B. 1984. Evolution of jaw mechanisms in ornithopod dinosaurs. *Advances in Anatomy Embryology and Cell Biology* 87:1–110.

Weishampel, D.B. and D.B. Norman. 1989. The evolution of occlusion and jaw mechanics in Late Paleozoic and Mesozoic herbivores. *Geological Society of America Special Paper* 238:87–100.

Wilkinson, D.M., E.G. Nisbet, and G.D. Ruxton. 2012. Could methane produced by sauropod dinosaurs have helped drive Mesozoic climate warmth? *Current Biology* 22:R292–R293.

Wolfe, J.A. 1979. Temperature parameters of humid to mesic forests of eastern Asia and relation to forests of other regions of the northern hemisphere and Australasia. *United States Geological Survey Professional Paper* 1106:1–37.

Wolfe, J.A. and G.R. Upchurch. 1987. North American nonmarine climate and vegetation during the Late Cretaceous. *Palaeogeography, Palaeoclimatology, Palaeoecology* 61:33–77.

Zhou, J., C.J. Poulsen, N. Rosenbloom, C. Shields, and B. Briegleb. 2012. Vegetation–climate interactions in the warm mid-Cretaceous. *Climate of the Past* 8:565–576.

# 9 The State of the Ancient Arctic

The Gunnison River of western Colorado is famous as a sport fishing destination for those interested in fly-fishing for various kinds of trout. It is also the river that flows through a magnificent canyon called Black Canyon of the Gunnison. The canyon, which can be thousands of feet deep, exposes some of the oldest rocks in North America, and in some places along the canyon it is deeper than it is wide. On top of these ancient rocks is a sequence of younger rock units that include rocks that contain dinosaur remains (Fiorillo and May, 1996). First established as a national monument in 1933 to preserve this natural wonder, the United States National Park Service later made this park a national park in 1999. Upriver from the park is the town of Gunnison.

In 1994, I started a 3-year paleontological survey of this park and the neighboring Curecanti National Recreation Area in western Colorado. Early in the project, park personnel recounted that close to the town of Gunnison the river had, through the natural course of things, moved. In other words, the river changed its main channel from a point close to town to one several hundred meters away from town. The result of this channel migration is that a very convenient destination for fishing guides was now no longer as convenient. The guide community attempted to apply pressure for the National Park Service to move the channel back to its old position. Because this channel shift technically occurred outside of both Black Canyon of the Gunnison National Park and Curecanti National Recreation Area, the National Park Service did not have to respond. The point of this anecdote is that there was a plea for action by some groups that carried with it an environmental modification and a significant financial burden. Would a better understanding of the geologic history of the river and its dynamics have avoided any of the contentious conversations within the community that followed a natural process of the river? Admittedly, this question is of local interest, but because there is only one variable in this change to the environment—the migration of the river channel—this story also illustrates how looking to the past helps us understand the future (Figure 9.1).

On a much broader and more complex scale, a casual examination of a variety of media today would show that the magnitude of environmental changes within the Arctic as the global climate changes is truly remarkable. The impacts of climate change are more profoundly felt in the Arctic than in the midlatitudes, and this phenomenon is termed Arctic amplification. The modern Arctic faces several environmental challenges, not all of which are independent of each other, not only in terms of physical changes but also changes in biological diversity. For example, Polar bears (*Ursus maritimus*) have become the poster child of Arctic ecosystem health, and organizations such as the Swiss-based International Union for Conservation of Nature have determined that global populations are in decline due to the pace of environmental change within the Arctic exceeding the pace at which these animals

207

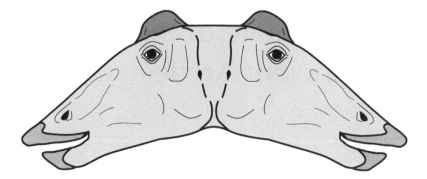

**FIGURE 9.1**   A dinosaurian version of Janus, or more true to the Latin spelling, *Ianus*—the two-faced Roman God. The two faces represent beginnings and endings, or looking back to look forward. In that sense, by looking back at the Cretaceous of the ancient high latitudes, I suggest that we may learn something about the future of a warming Arctic—hence, this paleontological reference to the ancient Roman God, perhaps more appropriately referenced within a dinosaurian context as *Ianusaurus*.

can adapt to it. Can the Arctic biota of the past help us understand the adaptability of life in an extreme environment?

There are so many different perspectives on what people mean by "the Arctic," and scale plays a role in the parameters of how we define the region. For example, at one scale one could argue that there are four ecoregions within the polar region: a polar domain, an ice cap division, a tundra division, and a subarctic division (Bailey, 1998). On a finer scale, one could suggest instead that there are some 20 ecoregions just within the state of Alaska (Gallant et al., 1995). The definitions for these divisions are not as important as the recognition that Alaska consists of a mosaic—a heterogeneous landscape.

Global climate change is a topic of intense concern, because it will have profound physical and biological effects on the world, and the scale of how we look at climate change can determine what we consider relevant to the discussion. Modeling the future of the Arctic suggests warming of several degrees over the next 100 years, while ice cores measure climate variability back several thousand years. Further, while most climate models point out an intensification of precipitation in the Arctic (e.g., Kusunoki et al., 2015), some models suggest the potential for greatly increased rainfall in the winter (Carter et al., 2000). Even in a warmer Arctic, some of that increased winter precipitation would likely come as snowfall, accumulate in layers, and be stored on the surface as snowpack, particularly in the higher elevations of the Brooks Range and other mountain ranges of the Arctic region. Such an increase in available snow for springtime melt would likely lead to an increase in significant early summer flooding down on the coastal plain.

The Arctic bonebeds that produced the thousands of dinosaur bones that form the basis of much of what we know about Arctic dinosaurs were formed under unique conditions and regional climate. These bonebeds also represent what these climate models suggest is ahead for this part of the Arctic. This ancient Arctic coastal plain was influenced by seasonally varying hydrologic processes, water levels, and by what appear to have been episodic floods. It is easy to imagine severe spring runoff events due to the paleogeographic juxtaposition of the towering ancestral Brooks Range, a narrow coastal plain, and the Arctic Ocean of the Cretaceous. Spring melting of high-altitude snowpack may sufficiently explain some or all of these bonebeds (Fiorillo et al., 2010; Flaig et al., 2014). This Cretaceous Arctic model illustrates the additional environmental impacts that would result from the precipitation predictions of some climate models for the future warming Arctic.

## DINOSAURS OF THE ANCIENT ARCTIC

Though Alaska assembled from enormous geologic blocks brought together by tectonic processes, some of which originated far from their present location, others of these blocks of land were near their current latitudinal position or higher. Cretaceous dinosaurs from Alaska were not only discovered in what are the high latitudes now, but they also lived in the high latitudes while they were alive.

As discussed throughout this work, over the course of the last few decades the study of the Cretaceous dinosaurs of Alaska and their ecosystem has opened the door to new research questions regarding the paleobiology of these fossil animals and plants as well as what we mean by the term "Arctic." Arctic dinosaur discoveries have two major points of significance. First, because these dinosaurs were year-round residents of the high latitudes, the study of Alaskan dinosaurs has shed new light on the paleobiology of these animals. It is clear from the growing fossil record that Arctic dinosaurs were well adapted to their environment and that they not only lived at the top of the world, but by exhibiting some of the patterns of behavior that we can see in modern animals, these ancient denizens of the North thrived in their respective habitats.

Second, the adaptations exhibited by these dinosaurs for living in the high latitudes contribute to our concept of the ecological pressures exerted by the physical environment of the Arctic. The seasonally extreme environment of the Cretaceous provided evolutionary opportunities for advantages for dinosaurs. The window in which high-quality food resources were available to herbivores was limited and when these foods were available, rapid growth occurred. During the long Arctic winter, high-quality food consumption diminished, as did growth. Patterns of relatively rapid growth followed by periods of slower growth are exactly what is observed in the cell structure of the Arctic dinosaurs *Edmontosaurus* and *Pachyrhinosaurus* (Erickson and Druckenmiller, 2011; Chinsamy et al., 2012). There exists a similar seasonal influence on patterns of growth observed in modern mammals ranging from northern ungulates to the Arctic ground squirrel (*Spermophilus undulates*; Guthrie, 1982). In modern mammals, this increased growth is attributed to many modern northern plants having greater concentration of nutrients per unit weight to compensate for such short growing seasons, a botanical pattern that may have been similar in the Cretaceous. Such adaptive similarities show convergence toward a basic framework for organismal responses under extreme ecological conditions.

Because the prey did not migrate, the predators similarly did not, and it is in these meat-eaters, specifically *Troodon* and *Nanuqsaurus*, that adaptations for a year-round polar existence are most recognizably expressed. For example, Alaskan individuals of the otherwise small theropod *Troodon* were shown to have achieved body sizes 50% larger than specimens from southern Alberta and Montana. The increased size was attributed to the larger eyes in proportion to body size of *Troodon* compared to other Arctic theropods, indicating that this dinosaur relied on sight. The larger eyes provided a competitive advantage in accessing more prey items in the low-light conditions of the ancient Arctic (Fiorillo, 2008a).

Conversely, the estimated skull length of *Nanuqsaurus hoglundi* is smaller than coeval and related tyrannosaurids from the lower latitudes. In contrast to troodontids, derived tyrannosaurs primarily relied on the sense of smell, rather than vision, and it seems reasonable to assume that the difference in emphasis on sight versus smell drove their distinctly different adaptive responses to the ancient polar world.

The increase in body size in one theropod taxon that in lower latitudes is much smaller, and the reduction in body size by a second theropod taxon that is part of a very large-bodied group of meat-eating dinosaurs in the lower latitudes, suggest that there may have been external ecological pressure for an optimal body size for predatory dinosaurs in the seasonal Cretaceous Arctic ecosystem of northern Alaska. This convergence in body size was dictated by the effective net biological productivity during the growing season. For *Nanuqsaurus hoglundi* the resource limits selected for smaller body size, while for the sympatric *Troodon* the adaptive advantage of larger eyes in the highly seasonal physical environment selected for larger body size. These two predatory dinosaur taxa, as well as the patterns of growth observed in the herbivorous *Edmontosaurus* and *Pachyrhinosaurus*, show that the dinosaurs of the ancient Arctic were well adapted to their environment. Though the Arctic climate may change, and those changes likely carry significant consequences, perhaps the insight gained from the fossil record is one of optimism; allowed enough time, life adapts even under extreme conditions.

If dinosaurs spent their winters in an environment with an extreme light regime, how did they survive? Even though the climate of Alaska was much warmer than it is today on average, the dinosaurs had to contend with long, dark winter seasons.

A decoupling of energy flow within an organism, as observed in modern reptiles (Shine, 2005), could be the key. Arctic dinosaurs may have effectively starved during the winter and reproduced during the short summer. Some dinosaurs were and are endotherms, but perhaps some were instead more like reptiles in their physiology—the large-bodied extreme of an ectothermic life history phenomenon. Rather than a near-complete metabolic shutdown for large herbivores during periods of dark when deciduous plants lost their leaves, delignified wood may have served as a low-grade food source for the plant eaters, as long as they could maintain body temperature high enough to function.

In addition, there is the issue of biogeography. That a land bridge connection occurred intermittently is likely responsible for the exchanges of a broad suite of terrestrial organisms between the two modern land masses of Asia and North America through the Tertiary and Quaternary is a well-established concept in Earth history. This same region, referred to as Beringia, is also the most likely route for dinosaur migrations between central Asia and western North America in the Cretaceous (Russell, 1993, Cifelli et al., 1997; Sereno, 2000; Fiorillo, 2008b; Zanno, 2010; Fiorillo and Adams, 2012). Such descriptive efforts in paleobiogeography are of course necessary in documenting the phenomenon, but attempting to understand the causal mechanisms for those biotic exchanges is arguably the more dynamic challenge.

The bonebeds that have yielded much of what we know about Arctic dinosaurs are spectacular in terms of abundance of fossil material and the lateral extent of the bonebeds. These bonebeds show that the large herbivorous dinosaurs, the hadrosaurs and the ceratopsians, lived in herds (Gangloff and Fiorillo, 2010; Fiorillo et al., 2010; Mori et al., 2016), and while predatory dinosaurs almost certainly wandered the landscape accordingly, it seems that the hadrosaurs and ceratopsians preferred different types of sub-habitat on the ancient Arctic coastal plain—the former preferring the wetter, poorly drained coastal lowlands, and the latter preferring the better drained upper coastal plain (Fiorillo et al., 2016).

Similarly, the incredible footprint record now known from around the state, and particularly Denali National Park, indicates distinct habitat preferences. In the well-studied Cretaceous rocks of Denali National Park, there is ample evidence that the physical landscape varied, as did the distribution of plants and animals yielding a heterogeneous landscape (Fiorillo et al., 2015). The physical aspects of these environments, such as coastal lowlands and coastal uplands, exist in the modern world. What is different are the inhabitants of these environments today compared to the deep past.

Thus, this understanding of habitat preferences and landscape heterogeneity and how they might change through time and space enlightens us on the factors influencing biotic response in this ancient high-latitude terrestrial ecosystem. Continued fine-tuning of our understanding of these details of these habitat preferences will illuminate us on critical aspects about biotic responses to climate change through deep geologic time.

The concept of Beringia, a geographic area within the high latitudes that encompasses far eastern Asia and northwestern North America, as suggested by David Hopkins (1967, 1996), was determined by the combination of the internal dynamics of the planet, which built the physical pathway, combined with external dynamics such as the atmosphere and climate, which allowed the pathway to be utilized. The origin of Beringia lies in the Cretaceous—Alaskan dinosaurs prove that—and the defining determinant is in the accretionary tectonic history of Alaska, a history responsible for the creation of the geographic landscape that was available for biotic exchanges between Asia and North America. Understanding habitat characteristics of this ancient ecosystem, the tectonic history, and the variations in climate and changes in climate are key components providing not only details of the timing of biogeographic events, but also detailed insights into the mechanisms that drove such events.

Within Cretaceous Beringia, *Nipponosaurus sachalinensis* is a crested hadrosaur dinosaur known from Sakhalin Island in the northern Pacific. This dinosaur had been thought to be closely related to the North American hadrosaurid *Hypacrosaurus* (Suzuki et al., 2004). A recent re-examination of this hadrosaur has shown the animal to instead be more closely related to Europe's hadrosaurids *Blasisaurus* and *Arenysaurus* (Takasaki et al., 2017). This evolutionary relationship suggests that *Nipponosaurus* is a dinosaur that migrated from Europe to far eastern Asia, rather than arrived in this region from North America (Takasaki et al., 2017). With the land bridge in place at that time, what fine-scale ecological components may have been in place during the Cretaceous that served as a biogeographic filter that precluded crested hadrosaurs from moving between Asia and North America?

During the Last Glacial Maximum, there existed an expansive ecosystem known as the mammoth steppe. This ecosystem had a cold, dry climate, with a vegetative landscape dominated by herbaceous plants. These plants supported large numbers of large herbivorous animals such as woolly mammoths (*Mammuthus primigenius*), bison (*Bison priscus*), and horses. These animals freely moved within this ecosystem between Eurasia and North America. As climate changed, the mammoth steppe deteriorated as warmer, wetter conditions took over, conditions that created wetlands in place of grasslands. Such environmental changes fragmented populations of animals, which ultimately led to extinctions. Similarly, we know that the terrestrial environments across the Alaskan Cretaceous were heterogeneous, so it is very likely that a sub-environment existed within this region that prevented crested hadrosaurs, like *Nipponosaurus*, from using the land bridge connecting Asia and North America. Speculating further, as the modern Arctic changes, do these examples foretell that existing populations of animals (and plants) will become increasingly fragmented populations?

"Ice Age" migrations between Asia and North America occurred as a function of specific available habitat (i.e., tundra–steppe conditions); understanding the nature of the Cretaceous habitat preferences for fossil vertebrates would serve to explain what controlled biotic migrations between the continents in deep time. Given the differences in climate between the Cretaceous of Alaska and Quaternary Beringia, further comparative study will likely provide insight into what is meant by ancient high-latitude ecosystems and perhaps even offer insight into the origins of what constitutes the Arctic. To borrow the question Hopkins (1996) posed for the geologically more recent understanding of Beringia, to what extent can we reconstruct ancient environments on or near the Beringian land bridge?

If we are successful in increasing our resolution in these ecosystem reconstructions, then we can contribute more broadly than just basic biogeography—these studies can also contribute to applied aspects of the ecological sciences. For example, conservation biology is one of the newest, and perhaps most directly relevant disciplines in biology with respect to the overall health of life on Earth. The discipline provides a long-term perspective on how communities change at various temporal and spatial scales, through the application of paleontology datasets, theories, and analytical tools as they relate to conservation issues—perhaps most importantly, issues related to species extinction. Advocates for this new perspective on biological science endorse a holistic rather than a reductionist approach to problem solving, with the hope that the discipline can inform science-based decision-making (e.g., Soulé, 1985; Meffe and Viderman, 1995; Groves et al., 2002). Within conservation biology, one of the main focuses is to recognize the underlying ecological processes that support primary productivity and biodiversity. How does the primary productivity compare across time planes, or from environment to environment? Vertebrate paleontology, by working in deeper time, contributes to the emerging discipline because it is the tool by which we can define what is meant by "natural" within ecosystems in terms of ecosystem composition and structure. The footprint record from the Lower Cantwell Formation of Denali National Park shows that some of these dinosaurs—specifically the hadrosaurs—lived in herds with multiple generations of individuals, demonstrating complex social structure and parental care given that the smallest individuals in the herd were too small to engage in long-distance migrations (Fiorillo et al., 2014; Fiorillo and Tykoski, 2016). That these complex multigenerational herds existed indicates that dinosaur thrived within their environments, and the presence of so many large-bodied herbivores across the region (Fiorillo et al., 2010, 2014, 2016; Flaig et al., 2014) suggests that primary productivity was high within this high-latitude ecosystem, albeit seasonally. So, beyond the biogeographic questions surrounding Beringia, given the sensitivity of Arctic terrestrial ecosystems to system changes, an enhanced understanding of these ancient Arctic environments and how the biota interacted with the environments will contribute greatly to society's goal of understanding fundamental questions about ecosystem processes.

As with any fascinating and challenging scientific project, we are just learning how to ask the questions, and it is clear from the results of these first few decades of work that the answers sought are not only satisfying for those interested in the paleobiology of those magnificent animals of the Mesozoic and the ecosystems in which they lived, but contribute to our understanding the world today and the impacts of how climates and ecosystems change. Such results and the need for more only illuminate that not only have we indeed entered "The Age of the Ancient Arctic," but what we are learning has far-reaching implications across a wide array of scientific disciplines.

## REFERENCES

Bailey, R.G. 1998. *Ecoregions: The Ecosystem Geography of Oceans and Continents.* New York, NY: Springer-Verlag.

Carter, T.R., M. Hulme, S. Crossley, M.G. Malyshev, M.G. New, M.E. Schlesinger, and H. Toumenvirta. 2000. *Climate Change in the 21st Century: Interim Characterizations Based on the New IPCC Emissions Scenarios.* The Finnish Environment 433. Helsinki, Finland: Finnish Environment Institute.

Chinsamy, A., D.B. Thomas, A.R. Tumarkin-Deratzian, and A.R. Fiorillo. 2012. Hadrosaurs were perennial polar residents. *The Anatomical Record* 295:610–614.

Cifelli, R.L., J.I. Kirkland, A. Weil, A.L. Deino, and B.J. Kowallis. 1997. High-precision $^{40}Ar/^{39}Ar$ geochronology and the advent of North America's Late Cretaceous terrestrial fauna. *Proceedings of the National Academy of Sciences of the United States of America* 94:11,163–11,167.

Erickson, G.M. and P.S. Druckenmiller. 2011. Longevity and growth rate estimates for a polar dinosaur: A *Pachyrhinosaurus* (Dinosauria: Neoceratopsia) specimen from the North Slope of Alaska showing a complete developmental record. *Historical Biology* 23:327–334.

Fiorillo, A.R. 2008a. On the occurrence of exceptionally large teeth of *Troodon* (Dinosauria: Saurischia) from the Late Cretaceous of northern Alaska. *Palaios* 23:322–328.

Fiorillo, A.R. 2008b. Cretaceous dinosaurs of Alaska: Implications for the origins of Beringia. In *The Terrane Puzzle: New Perspectives on Paleontology and Stratigraphy from the North American Cordillera*, eds. R.B. Blodgett and G. Stanley, 313–326. Geological Society of America Special Paper 442.

Fiorillo, A.R. and T.L. Adams. 2012. A therizinosaur track from the Lower Cantwell Formation (Upper Cretaceous) of Denali National Park, Alaska. *Palaios* 27:395–400.

Fiorillo, A.R., S.T. Hasiotis, and Y. Kobayashi. 2014. Herd structure in Late Cretaceous polar dinosaurs: A remarkable new dinosaur tracksite, Denali National Park, Alaska, USA. *Geology* 42:719–722.

Fiorillo, A.R., Y. Kobayashi, P.J. McCarthy, T.C. Wright, and C.S. Tomsich. 2015. Reports of pterosaur tracks from the Lower Cantwell Formation (Campanian-Maastrichtian) of Denali National Park, Alaska, USA, with comments about landscape heterogeneity and habitat preference. *Historical Biology* 27:672–683.

Fiorillo, A.R. and C.L. May. 1996. Preliminary report on the taphonomy and depositional setting of a new dinosaur locality in the Morrison Formation (Brushy Basin Member) of Curecanti National Recreation Area, Colorado. *Museum of Northern Arizona Bulletin* 60:555–561.

Fiorillo, A.R., P.J. McCarthy, and P.P. Flaig. 2016. A multi-disciplinary perspective on habitat preferences among dinosaurs in a Cretaceous Arctic greenhouse world, North Slope, Alaska (Prince Creek Formation: Lower Maastrichtian). *Palaeogeography, Palaeoclimatology, Palaeoecology* 441:377–389.

Fiorillo, A.R., P.J. McCarthy, P.P. Flaig, E. Brandlen, D.W. Norton, P. Zippi, L. Jacobs, and R.A. Gangloff. 2010. Paleontology and paleoenvironmental interpretation of the Kikak-Tegoseak Quarry (Prince Creek Formation: Late Cretaceous), northern Alaska: A multi-disciplinary study of a high-latitude ceratopsian dinosaur bonebed. In *New Perspectives on Horned Dinosaurs*, eds. M.J. Ryan, B.J. Chinnery-Allgeier, and D.A. Eberth, 456–477. Bloomington, IN: Indiana University Press.

Fiorillo, A.R. and R.S. Tykoski. 2016. Small hadrosaur manus and pes tracks from the lower Cantwell Formation (Upper Cretaceous), Denali National Park, Alaska: Implications for locomotion in juvenile hadrosaurs. *Palaios* 31:479–482.

Flaig, P.P., A.R. Fiorillo, and P.J. McCarthy. 2014. Dinosaur-bearing hyperconcentrated flows of Cretaceous Arctic Alaska: Recurring catastrophic event beds on a distal paleopolar coastal plain. *Palaios* 29:594–611.

Gallant, A.L., E.F. Binnian, J.M. Omernik, and M.B. Shasby. 1995. Ecoregions of Alaska. *United States Geological Survey Professional Paper* 1567:1–73.

Gangloff, R.A. and A.R. Fiorillo. 2010. Taphonomy and paleoecology of a bonebed from the Prince Creek Formation, North Slope, Alaska. *Palaios* 25:299–317.

Groves, C.R., D.B. Jensen, L.L. Valutis, K.H. Redford, M.L. Shaffer, J.M. Scott, J.V. Baumgartner, J.V. Higgins, M.W. Beck, and M.G. Anderson. 2002. Planning for biodiversity conservation: Putting conservation science into practice. *BioScience* 52:499–512.

Guthrie, R.D. 1982. Mammals of the mammoth steppe as paleoenvironmental indicators. In *Paleoecology of Beringia*, eds. D.M. Hopkins, J.V. Matthews, Jr., C.E. Schweger, and S.B. Young, 307–326. New York, NY: Academic Press, Inc.

Hopkins, D.M. (ed.). 1967. *The Bering Land Bridge*. Stanford, CA: Stanford University Press.

Hopkins, D.M. 1996. Introduction: The concept of Beringia, In *American Beginnings: The Prehistory and Palaeoecology of Beringia*, ed. F.H. West, xvii–xxi. Chicago, IL: University of Chicago Press.

Kusunoki, S., R. Mizuta, and M. Hosaka. 2015. Future changes in precipitation intensity over the Arctic projected by a global atmospheric model with a 60-km grid size. *Polar Science* 9:277–292.

Meffe, G.K. and S. Viederman. 1995. Combining science and policy in conservation biology. *Wildlife Society Bulletin* 23:327–332.

Mori, H., P.S. Druckenmiller, and G.M. Erickson. 2016. A new Arctic hadrosaurid from the Prince Creek Formation (lower Maastrichtian) of northern Alaska. *Acta Palaeontologica Polonica* 61:15–32.

Russell, D.A. 1993. The role of central Asia in dinosaurian biogeography. *Canadian Journal of Earth Sciences* 30:2002–2012.

Sereno, P.C. 2000. The fossil record, systematics and evolution of pachycephalosaurs and ceratopsians from Asia. In *The Age of Dinosaurs in Russia and Mongolia*, eds. M.J. Benton, M.A. Shishkin, D.M. Unwin, and E.N. Kurochkin, 480–516. Cambridge, UK: Cambridge University Press.

Shine, R. 2005. Life-history evolution in reptiles. *Annual Review of Ecology, Evolution and Systematics* 36:23–46.

Soulé, M.E. 1985. What is conservation biology? *BioScience* 35:727–734.

Suzuki D., D.B. Weishampel, and N. Minoura. 2004. *Nipponosaurus sachalinensis* (Dinosauria; Ornithopoda): Anatomy and systematic position within Hadrosauridae. *Journal of Vertebrate Paleontology* 24:145–164.

Takasaki, R., K. Chiba, Y. Kobayashi, P.J. Currie, and A.R. Fiorillo. 2017. Reanalysis of the phylogenetic status of *Nipponosaurus sachalinensis* (Ornithopoda: Dinosauria) from the Late Cretaceous of Southern Sakhalin. *Historical Biology* 5:1–18. doi:10.1080/08912963.2017.1317766.

Zanno, L.E. 2010. A taxonomic and phylogenetic re-evaluation of Therizinosauria (Dinosauria: Maniraptora). *Journal of Systematic Palaeontology* 8:503–543.

# Index

T - #0136 - 111024 - C240 - 234/156/11 - PB - 9780367657444 - Gloss Lamination